MW01491960

Stories from the
HEARTLAND

by

Max Armstrong
One of America's Favorite Farm Broadcasters

BANTRY BAY PUBLISHING

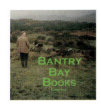

Stories from the Heartland
by
Max Armstrong
Copyright © 2015 Max Armstrong
All rights reserved.

Correspondence to the publisher should be by email:
bantrybaypublishing@gmail.com,
or by telephone: (312) 496-3677

Printed in the United States of America
at Lake Book Manufacturing, Melrose Park, Illinois

ISBN 978-0-9850-6738-0

Library of Congress Cataloging— Publication Data has been applied for.

Photos and images are the property of Max Armstrong, unless otherwise indicated.

The drawing of a microphone used at the end of chapters was drawn by Lisa Armstrong, Max and Linda's daughter.

The Farmer's Creed (pp. 22-23) is courtesy CNH Industrial America LLC/New Holland

The photo behind the Farmer's Creed is courtesy of Terese Husnik of Vinton, Iowa, who photographed her daughter, Heidi Primrose, Heidi's husband Brad, and their three children, Lexi, Livi and Will on their cattle and corn farm in Shellsburg, Iowa.

Table of Contents

Dedicated
to the men and women of the Heartland.
Without them, none of what I do would be possible.

The Jim and Stella Fay Armstrong farm near Owensville, Indiana

Introduction

Every now and then over the years, I've wondered how my life would have been different had I not grown up on Jim and Stella Fay Armstrong's Indiana corn and soybean farm. I always seem to come to the same conclusion: Whatever it would have been, my life would not have turned out as rich and as rewarding. I don't see how it could possibly be. I truly thank God each day for the parents we had and the manner in which we grew up.

Many wise and thoughtful people have written volumes about the lessons learned from country life. You've heard it all before — heck, you've probably lived it. You've learned about being responsible for maintaining another creature's life, nurturing plants from seed to harvest, learning — often the hard way — respect for Mother Nature and her unpredictability, and never taking for granted the opportunities that are right in front of us, although they may be obscured by dust and mud at times.

This book is my modest attempt at honoring that lifestyle and the people of the heartland who do the hard work every day to keep food on our tables. My occupation in front of a microphone and TV camera has given me the opportunity to set foot in places I could never have imagined, and I met some remarkable folks along the way. Their stories, and my observations and photos, are in the pages that follow.

I must share with you, though, that after nearly 40 years of originating broadcasts from every state in America and more than 30 different countries, seeing the most awesome mountains, lakes, sunsets and skylines on this planet, I will happily tell anyone, "Give me the country life."

Max Armstrong

The Armstrong Family "Super H" and "Super M"

Signs like this should be erected at the edge of towns all across America. Maybe farmer-funded commodity check-off programs could help foot the bill and county Farm Bureaus could implement it. Or maybe a major agribusiness could step up to help make it happen.

At the Vintage Wings and Wheels Museum in Poplar Grove, Illinois, with Orion and our vintage wheels: His family's 1939 Farmall F-20 and our family's 1962 Farmall 560.

Foreword

As I was thinking about what I'd say about Max, it dawned on me just how long I've known him. When I met Max, I was still a young man, in my 40s! I first heard him when he was working for the Illinois Farm Bureau, producing farm reports that went to radio stations across the state. He was good, and I told my assistant, Lottie Kearns, that if we ever had an opening, I'd be contacting young Max Armstrong.

That happened in 1977. I tried to hire Max, but, to my surprise, he didn't seem all that interested. He had settled in to the Towanda, Illinois, area where he was on the volunteer fire department. Eventually, I got him, and I've had a lot of fun with the story that he struggled to decide between a coveted job at the legendary WGN and an unpaid job as a firefighter.

Not only did he become one of the best farm broadcasters in WGN's history and in the nation, Max became a business partner and a valued friend. We were together for 31 years at WGN Radio and TV until Max received an offer from the Penton Farm Progress companies that he couldn't refuse. We're still on the air together though, each week on *This Week in AgriBusiness*, which airs on RFD-TV and several TV stations throughout the country, and each Saturday on WGN's *The Saturday Morning Show*.

Max is a master of the art of storytelling, and I know you'll enjoy the stories and photos he shares in this book.

Max continues to be a leading voice for agriculture. The men and women who feed us, clothe us, and put fuel in our tanks and roofs over our heads are lucky to have him as their advocate. And I'm lucky to have Max as a friend and partner.

Orion "Big O" Samuelson

Jim and Stella Fay Armstrong

— *Chapter One* —

Farmers

*T*hey are amazing. Their ingenuity and entrepreneurial spirit, their creativity, optimism, and compassion for others has always made me proud. I hope we don't lose that.

There isn't enough praise for the men and women who have chosen the farm life. It is difficult and full of danger. There are often brutal conditions, long hours and incredible risk, but the rewards are tremendous and follow you throughout your life.

I'm so fortunate to know so many and to have been in their homes and on their farms and ranches, sitting in their cabs with them. You realize there are a lot of things they struggle with. Of course, the big one is weather, but they struggle with the government, too. And they're not exempt from all the challenges that any family has. Rural Americans, unfortunately, suffer some of the same foibles as those in big cities.

Yet, there's so much to admire about these people. One quality is the spirit of helping each other in times of need, which we've seen expressed in so many wonderful ways. I hope the new generation coming on retains that quality.

My friends in agriculture are such a great source of knowledge for me. Many of them may say I still have a lot to learn, but the people I've met through the tractors, farm shows, through agriculture leadership and farm organizations, have given me an education I could never pay for.

I've been at it long enough that I now meet the sons, daughters — heck, even grandchildren — of the people I interviewed 30 years ago. I hear, "My granddad was president of the Corn Growers," or "of the Pork Producers." It's neat to run into the new generation of farmers.

They are people of great faith in most instances. They often have to draw on that faith because of challenges that can come in a variety of ways, usually from Mother Nature. Like the spring and summer of 2015, when farmers from Western Ohio through Eastern Illinois watched helplessly day after day as their fields were hit with storms. For not just young farmers, but for their parents, this relentless rain was something they hadn't seen in their careers.

Of course, we've seen drought, too, and we had a devastating one in 2012. Losing a crop, whether watching it burn up for lack of rain, or rot because of too much, is difficult for a farmer to watch. It's understandable when producers become emotional about it because it's their livelihood. If you lose livestock, as Midwest poultry producers did with the bird flu — called by some the worst such disaster in U.S. history — you don't just rebuild overnight. Although there are ways to protect some of the investment through insurance, you're not made whole.

Another thing that strikes me about farmers is how much they care for the land where they live. So many times, I've often ridden with farmers past their fields and they'll point out improvements that they've made. Maybe it's structures they've put up, or drainage tiles they've added. And they tell how they've worked to make sure what they're doing on their land isn't impacting others. For example, a farmer might work through a local drainage board to keep the ditches open, not only so they'll drain well but also not wash out the bridges the school bus goes over. They really care for the land and want to make it better. Farmers often use the term stewardship, and it's not a throwaway phrase. They really mean it. They worry and agonize over the land and want to leave it better than they got it.

Many lessons are learned on the farm. A huge one is work ethic, which carries over into all parts of life, and employers recognize that when they hire someone with farm experience, they're getting someone special. In the 22 years I served on the Board of Fire Commissioners, we'd sometimes have a farm kid applying for a job as a full-time firefighter. When you looked at his resume and saw that he had livestock, you knew that 99% of the time, this was a young person who had to be out there every day, no matter the weather, no matter how long it took, to care for those animals. They also knew this is someone with "know-how," with mechanical skills and an ability to fix things. A young person on a farm learns so much about how to get along in life.

We're in a very competitive era of agriculture now. Neighbors are often competitors. As farmers continue to expand the acreage they farm to benefit from the economies of scale, they are often bidding against each other on the cash rental rates with the landowners. The pressure has been dialed up as farms have gotten larger to survive. It can get intense.

Some of the pressure is because agriculture is under attack, often from within. Take the organic movement, for example. Organic producers and proponents say organic farming is the best for the environment, healthier for humans, and is, simply, the only way to go. A lot of smoke and mirrors are used to attack traditional agriculture, which, if we're going to continue feeding the world, is the way we *have* to go. There's not a shred of peer-reviewed, third-party highly regarded evidence that organic is better

THIS PHOTO OF ME AS A TEENAGER brings back memories of long, hot days guiding the "Super M" around the Wabash Bottoms. But as the sizzling sun that had tortured me all day settled behind the tree line, it was time to head to the house. There just were not many things that felt better than pulling the implement out of the ground, easing the tractor out onto the blacktop, and shifting into "road gear." Heading up the road at a blazing 15 miles per hour with the blast of cool evening air taking some of the sunburn sting away, it felt so good to have done a day's work. All these years later, hardly anything has ever looked as pretty as coming around the curve down by the barn and catching the first glimpse of the warm, yellow glow in the darkness from Mom's kitchen window, and knowing what was soon to be on my plate.

for you. None. Nowhere. And yet, there is that perception that if it's organic, it's good for you.

Is there room for organic? Of course, and if you feel better from consuming organic, then absolutely buy it. And I encourage a producer to get as much as he/she possibly can for producing that organic product. I wish those farmers well, but organic farming can't produce enough food for the seven billion people on earth. And when aspersions are cast on our basic, traditionally-produced crops, that's where I part company with some of the organic crowd.

I cruise through Whole Foods sometimes and presumably, these people are shopping there because they are concerned about purity, quality and want to buy the healthiest food they can. Well, there are nearly always sample bowls. Have you ever watched people put their dirty hands in there, touching other chips or whatever the food is before taking theirs? And their sleeves are touching the side of the sampling bowl going in or coming out. And these are people concerned about what they're putting in their mouths?

Piggybacking on the organic issue is the debate about what "local" means. It's seen as trendy and somehow anti-Big Ag if you eat "locally-sourced" food. But if you go to a farmer's market in Chicago and you're expecting to buy local produce, you're probably thinking it is no farther away than northwest Indiana or southwest Michigan. But some of the "local" farmers are from farms more than 250 miles away. You also have to consider what we are capable of growing locally. The Midwest has a limited growing season, and conditions aren't conducive to every type of food. If we didn't bring food in from other parts of the United States, and the world, we would starve.

Our traditional food and processing system is the envy of the world.

And yet you've got people saying, "If it's local, it's better." "If it's organic, it's better." "If it's non-GMO, it's better." "If it's not produced on a corporate, 'factory' farm, it's better." I'm not really sure what a "factory" farm is, but I generally like what comes out of a factory because it's uniform and of good quality. Farms that are large, integrated operations turn out a superb product.

Fortunately, traditional agriculture isn't letting these phony claims go unanswered. I can't tell you how proud I am of farm folks for standing up, crying foul, and telling their story on social media. They are posting photos, videos and stories on Facebook, Twitter, Instagram, Snapchat, Pinterest and other platforms to show and tell what they're doing on their farms. A lot of farmers are embracing this, especially women. It's neat to see this and even though there are many new ways and places to tell ag's story, it is far from a new issue. Orion was espousing this long before I came on the air nearly 40 years ago. Today, every farmer realizes they need to tell their story, and it's easier than ever to do so.

Pressure is nothing new for farmers who've been at it for a while, those who might have a little snow on the roof. They've lived for decades with ever-rising costs of doing business, unpredictable commodity values that are dictated by people who've never set foot on a farm, and, of course, the weather. They are perennial gamblers, never knowing whether the expensive seed they sowed in the expensive land with the expensive planters will ever produce a penny of value. But they're used to it. For the young farmer, perhaps fresh out of the University of Illinois or Purdue, this is a whole new ball game. Luckily, they are probably better positioned to handle this than their parents were because college taught them all the tools of risk management. And those who have a parent as a partner in the business to help guide them are even better positioned. I always tell young farmers that one of the best classrooms in the world is at the knees of a grandparent. Sure, they may not know about social media, don't know what a tweet is, but they have endured many difficult and challenging times. They can put things into perspective. I tell young people, "If you've got a grandparent, sit down, ask, and listen. Do what Max Armstrong did NOT do! Listen to them, and <u>record</u> them if you get a chance."

I don't know of anyone who captured the rewards that farming and ranching has to offer better than the ad agency employee who wrote "The Farmer's Creed," which I posted on the following two pages. The original version was credited to Frank I. Mann, an Illinois corn farmer who reportedly wrote it about 100 years ago, but it was updated and revised in 1995 by an ad agency for a New Holland ad.

I hope to learn the name of the author some day, but for now I'll just say thanks, and hope you enjoy it as much as I do.

I believe ...

Farmer's Creed

I believe a man's greatest possession is his dignity and that no calling bestows this more abundantly than farming.

I believe hard work and honest sweat are the building blocks of a person's character.

I believe that farming, despite its hardships and disappointments, is the most honest and honorable way a man can spend his days on this earth.

I believe farming nurtures the close family ties that make life rich in ways money can't buy.

I believe my children are learning values that will last a lifetime and can be learned in no other way.

I believe farming provides education for life and that no other occupation teaches so much about birth, growth, and maturity in such a variety of ways.

I believe many of the best things in life are indeed free: the splendor of a sunrise, the rapture of wide open spaces, and the exhilarating sight of your land greening each spring.

I believe that true happiness comes from watching your crops ripen in the field, your children grow tall in the sun, and your whole family feeling the pride that springs from their shared experience.

I believe that by my toil I am giving more to the world than I am taking from it; an honor that does not come to all men.

I believe my life will be measured ultimately by what I have done for my fellow man, and by this standard I fear no judgment.

I believe when a man grows old and sums up his days, he should be able to stand tall and feel pride in the life he's lived.

I believe in farming because it makes all this possible.

Stella Fay "Switchblade" Armstrong

Chapter Two

Mom

As I was cleaning out my mother's nightstand after she died, I found the note, below, in her handwriting. I found that note right next to her switchblade. I picked up the knife, hit the button and a razor-sharp blade flipped out. Oh my goodness! Mom and Dad had an issue with some character breaking a basement window, and Mom was ready for them. And she meant business!

Mom was a nurse, but discontinued her career when she married Dad. She did her training in Chicago at the Lying-In Hospital, which is now the University of Chicago Medical Center. She loved the career and the experiences, and also loved Chicago. As a matter of fact, both my

> Politicians and diapers should be changed often – and for the same reason.

parents loved and enjoyed the city.

Back on the farm, Mom did what farm women had done for decades: worked alongside my dad on the farm by taking care of us. She prepared the meals, ran errands, made sure the tanks had fuel, and if a truck needed to be moved, she moved it. We always had a huge garden and Mom did all the work. I can tell you that after a long day on a tractor, I couldn't get to her table fast enough. I knew that there would be some of the tastiest pork and gravy, flavorful garden sweet corn, luscious home-canned beets, soft homemade biscuits, and the richest blackberry cobbler a farm boy had ever known. My "bottomless pit" needed filling, and Mom knew just how to do it. There were not many things that she seemed to enjoy more.

Things have changed, of course. While Mom was like the guy who

A few of you knew her, and in those years before Alzheimer's, Mom was both sweet and tough. But I never expected to find this wicked switchblade knife in her nightstand. The lesson here, boys and girls, is be kind to your grandma. You just never know what she might be packin'!

used to be on *Captain Kangaroo,* keeping all the plates spinning at one time, today's farm women are spinning a lot more plates. It's incredible to see what farm women do! Unfortunately, they're often the unsung heroes who don't get nearly enough credit.

Many of them help market crops successfully. Many care for live-stock, sometimes better than their male counterparts. And, as we all know, women are better communicators, and they use that skill to tell agriculture's story to the rest of the world. Many of them have set up their farms for agri-tourism, developing new streams of revenue. Visits by school groups are common, with many farm women involved in the "Ag in the Classroom" program.

Monsanto honors the Farm Woman of the Year, and I was in St. Louis one year to interiew the finalists. Their stories were really remarkable, and what most impressed me was how some of them came to agriculture by accident, never intending to be a part of it. Usually, that's because they met their husbands in college and fell in love not only with them, but with the farm life. They've often brought a new perspective and enthusiasm to the farm and are often the ones leading the way as ag tells its story. I'd like to think that Mom, were she a farm woman today, would be very involved in spreading the good word about agriculture.

My mother had Alzheimer's. It's a struggle that far too many families deal with. But even in those dark days when Mom was losing herself, there were lighter moments. Around Labor Day, in the last year of her life, she fell, hit her head and didn't talk after that. Her roommate sure did, though. She was very loud, and you could hear her all the way down the hallway. Around Thanksgiving, I was in their room at the nursing home and I was rubbing Mom's back. The other lady kept yelling, for everyone in the building to hear, "Hey, come over here and give me some of that!

Come over here for me!" Finally, Mom looked at her roommate, and in as distinct and clear a sentence that ever came from her lips, she said, "Would you please shut up?" Those were the last words I heard my mother say.

A phenomenon I observed while sitting with Mom and Dad at the nursing home church service was truly amazing. Many of the patients were struggling to speak, and it was difficult for them to read. Yet when it came time for the service, as I'm struggling with the hymn book, all of these older folks were mouthing the words of the hymns from heart. So many other things had left them, but those lyrics were still there.

Mom passed away in 2005, just three days shy of her 87th birthday. She and Dad, married 60 years, are buried together in Gibson County, Indiana. I sure miss being able to talk with her.

I am reminded of a story about Paul "Bear" Bryant, the legendary coach of the University of Alabama football team. He did a television commercial for Bell South, one of the regional "Baby Bell" companies before "Ma Bell" was broken up. Sitting at his office desk, he looked into the camera and said:

"One of the first things we tell our players is to stay in touch with their families. When our freshmen first arrive, we ask them to write a postcard home, right then. You know, we keep them pretty busy, but they always have time to pick up the phone and call. It's really important to keep in touch. Have you called your mama today? I sure wish I could call mine."

Amen, sir, amen.

*I might as well get this confession out of the
way: I wasn't always a red tractor boy.*

Jim Armstrong on the "Super H" before it was restored

— Chapter Three —

Dad

My dad was part of the "Greatest Generation," a description I believe Tom Brokaw is credited with, and I can't think of a better term to describe and honor those folks. In my case, and I know this to be true for many other people of my generation, the hardships and sacrifices made by our parents — unimaginable by today's standards — made all of the gains made by the Baby Boomer generation possible.

James Richard Armstrong was a patriotic American. He was a strong supporter of our country and the flag. Conservative, to be sure. He was a sharecropping dirt farmer, and I don't say that in a demeaning way, but as a way to help you visualize the challenges that he and Mom went through to support us. During the Depression, my grandfather had to split up the family while he went to Michigan to work in the auto plants. He kept a daughter and a son with him, and parceled out the other three kids to aunts and uncles in Indiana. My father lived with an uncle who was a farmer, and Dad continued to farm with him into his adult life.

It was a lean existence as I was growing up, but Dad and Mom worked hard and did what they had to do. I remember riding with Dad in a truck loaded with grain, and as we were crossing the scales of the elevator,

Jim Armstrong on his 1953 Farmall "Super H." This was the winter of 1986-87 at our Owensville, Indiana farm. He would have been 70, and this was about eight years before the "Super H" moved to Illinois and got a facelift.

he told the elevator guy how that load was to be divided up. The owner and my uncle got a lot, but I was surprised when I saw how little was left for Dad. I recall thinking that it sure didn't go very far toward supporting our family. But it was something.

He also supported us with money he earned fixing roofs. Dad developed quite a reputation for fixing leaks. My brother, Steve, and I got to climb up and help him a lot, and we always enjoyed a good view of what was happening on Main Street!

He was a very caring father and was concerned about our futures. Dad told us that farming is difficult and challenging in many different ways, and he encouraged us to broaden our horizons. "Keep your options open," he said. Farming had been good for our family, but, "take a look at all the opportunities out there." We did, and neither one of us wound up farming for a living. Steve is retired from the food industry where he was a vice-president of Human Resources for a large national brand. Today, he is a personal financial advisor, and has looked after his little brother much longer than he was obligated to. Both of us were fortunate enough to go to Purdue. We had minor scholarships that, even back then, didn't add up to much. Dad and Mom came up with the rest.

Sometimes when I'm on Dad's old "Super H," I think of those family tractors we love to collect, restore and admire, and how they aren't just hunks of iron. Those tractors enabled our parents to put a roof over our heads, feed and clothe us, and also to give many of us an opportunity to go away to college.

Like many Midwesterners, Mom and Dad really enjoyed listening to WGN. It was 300 miles away and the signal could be a little scratchy during the day, but thanks to the sky wave — the skip — WGN came in loud and clear after dark when atmospheric conditions were right. When I went to work with Orion in the late '70s and was doing a lot of TV, Mom and Dad wanted to be able to see their son on WGN-TV, but Channel 9's signal didn't reach that far. This was long before satellite dishes came along. Cable TV had just come to Owensville, but our farm wasn't in the town limits. So Mom and Dad went before the town board and asked if they could get cable run the quarter of a mile out to the farm, and the board said, yes, it could do that, if the farmhouse was in the town limits. Owensville annexed the farmhouse, not the rest of the farm, because Mom and Dad wanted to be able to see me on WGN-TV!

As Dad got older, he agonized over getting rid of his Shorthorn cattle. He really enjoyed them and he took care of them so well, they were almost like pets. When he was in his late 70s, Dad was on the board of the Rural Electric Cooperative, and had been chairman for a few years. Whenever he would have to go to Indianapolis for a board meeting, he would get up early to feed his Shorthorns. One morning, one of them bumped him and knocked him down. Dad decided right then and there to step it back a little bit. He got rid of the cattle and got off the Electric Board, as well.

I do admire people who get involved in associations, but there is a cost for the good they do, especially to those who rise to the leadership level. They give up personal time and often can't be at their kid's events. They benefit from it immensely because of the network, acquaintances and knowledge they gain. But they have to have somebody rock-solid back home, whether it be the spouse, brother or hired help to keep the operations running while they attend meetings in the state capital or in Washington, D.C. But they do miss some things back home.

One thing I really admired about my folks was that Dad knew when to shut off the combine. I don't ever remember being at a ball game — and I played basketball, baseball and football — when my folks weren't sitting in the stands. Some of the other farmer parents weren't always there and I don't demean them. They had work to tend to. But for Dad, the storm clouds may have been on the horizon and he may have needed to harvest and get a crop in right away, but he turned off the combine and came to town to watch me play in a game. They were always there.

Dad was always doing things for other people and not expecting anything in return. There were ministers at our church who weren't making much in the way of compensation and I vividly recall Dad leaving some cash and clothing for them, to help them along a little bit. You find that a

Jim and Stella Fay never missed a chance to go to the field and watch this kid play ball.

Three generations of Armstrongs: Dad, Kristi and me.

lot in small communities, and as Mom and Dad got older, we were lucky to have Kenny Dillon, a proud Korean War veteran and member of the community, who looked after them when I couldn't be there.

Dad helped a lot of relatives, mainly my uncles and spinster aunts. One of my favorite stories is about Dad's sister, Marjorie. He spent a lot of years looking after her. She was large, diabetic, and had many health concerns that finally caught up with her. As we were standing by the casket at her funeral, Dad leaned over to me and said, "Well, at least your Aunt Marjorie isn't going to be canceling out my vote anymore." She had been raised by a different aunt and uncle and they were of another political persuasion.

At Dad's funeral, I was surprised to learn that he was an inspiration to a lot of the younger guys in the community. You're at your father's funeral and guys come through that you didn't realize had much contact with your father, and you hear of what he had done for them or how he had inspired them in some way, well, it's neat to hear that.

We have a little bit of acreage left, but most of the farm was sold off to younger operators throughout the years. It's less than ten acres and the farmhouse is still part of the town of Owensville.

ON WHAT WOULD HAVE been Dad's 99th birthday, I came across his pocket knives. Daddy always carried one and made it clear that they would serve you well on the farm for so many applications. When I posted it on Facebook, I dressed it up a bit, telling the story that my dad used a knife to castrate hogs in the morning and offered me a slice of melon in the afternoon. I asked, "Did you wash that knife?" He said, with a grin, "Of course I did!"

The responses from Facebook friends showed that there's a lot of nostalgia involving knives; memories of dads and granddads using them for not only barnyard work, but splinter removal, fingernail cleaning, food slicing and more.

Back in fifth grade, most boys carried knives to school. We played mumbley-peg, flipping the knife perilously close to the legs of a companion across the way. Can you imagine that happening today?

IN THE BACKGROUND of the photo above of Dad and me, right above my head, you'll see what were once, before my time, the Armstrong family restroom facilities. It was a three-holer. I never asked Dad why there were three, and I certainly never saw them occupied at the same time.

And how about our "free-range" backyard flock of chickens? Yep, we were cutting-edge some five decades before it became the rage and we knew we could charge 20 gazillion dollars for a free-range chicken who's pecking at his poop. So when you're dining out and you see free-range on the menu, just keep in mind what that really means.

This photo was taken in 1998. Lisa, to my right, was eight, and Kristi, to Linda's left, was 12. Mom passed away in 2005; Dad, in 2008.

Father's Day 2012 brought both girls home to check up on the old fella. Lisa, on my right, had just graduated from DePaul University. Kristi got her degree from Purdue in 2008.

You didn't think I would pass up the chance to show you the newest Tractor Girl, granddaughter Miss Ellie, did you?

— *Chapter Four* —

The Armstrongs

T he current crop of Armstrongs got their start after Linda and I met on a blind date in Bloomington, Illinois. I was working for the Illinois Farm Bureau and a buddy I worked with had a girl-friend who had a first cousin they thought I should meet. Linda had heard me on the radio and based on my voice, she thought I was quite a bit older. Luckily, she took a chance and agreed to the date anyway.

Linda really loved my parents and did so much for them. I felt, in many regards, that they were her parents, too. Linda was a dental assistant by training and that's what she did for her forty-year career, during which, I might add, she worked for only three different dentists.

Linda has been a tremendous companion and partner, and the best supporter I could hope for. Frankly, she did a lot of the heavy lifting through the years in raising our girls, Kristi and Lisa. My job, not only the early hours and long days at WGN, but the frequent travel to visit farmers across the Midwest and around the world, meant that I was gone a lot.

Kristi, our older daughter, married a Southern lawyer. Yes, a law-yer, and I actually like Andy ... REALLY like him! In 2015, Kristi and Andy added a branch, or at least a twig, to the family tree with baby Ellie, who is providing a type of joy and satisfaction I haven't known before. Someone

Lisa and Kristi came home to celebrate Mother's Day 2015 with Linda, along with Jack, Lucy, and a photo bomber.

thinking it's the other way around.

Daughter Lisa is a graphic designer for a large Chicago real estate firm, and I'm grateful to her for letting us use her drawing of an old microphone as an ornament at the end of each chapter.

I met many of the people whose stories I share in the course of my business, not theirs. That was not the case with Jon Muraskas. Linda and I met him because of a baby boy born with big challenges. Grant Armstrong was a Trisomy 18 baby. He had an extra chromosome, an 18th. The result, we learned quickly, is not good. A child born with heart, breathing and feeding problems cannot expect a long life.

The guy who delivered this jarring news was Dr. Jonathan Muraskas. Just hours after Grant arrived in this world, he and Linda were transported to the neonatal intensive care unit supervised by Dr. Muraskas. I was handed a map and was told that I could go see them later. I can honestly say I had never felt more alone in my life, not knowing what challenges lay ahead of us. It is the kind of fear and confusion that thousands of new parents face every day, but not all are as fortunate to have a guy the caliber of Jon Muraskas caring for them.

To be sure, this nation is blessed with legions of dedicated medical professionals and magnificent state-of-the-art facilities. While we are repeatedly astounded by the cost of all of this, we should be thankful for so many people who generally know their trade well. All too often, though, their shortcoming is empathy. What many lack, by my assessment, is the ability to understand and appreciate the emotions that patients, and the patients' loved ones, are undergoing, and being able to communicate with them in a corresponding manner. How often have we heard about, or personally experienced, stunning, emotionally-devastating news delivered in a matter-of-fact, mechanical way? Then the patient is told to make sure their bill is in order on the way out, make the next appointment, and, "We'll see you in a few days."

With Grant, we were scared to death, but we also felt we, and our little boy, were in the best place we could have possibly been. I am more convinced of that today than ever, thanks to Jon Muraskas, who was one of those people we felt we had always known.

Shown at right, Dr. Muraskas practices in Maywood, Illinois, and is affiliated with multiple hospitals in the area. His dad was a physician, and Jon's brother is, too. Jon received his medical degree from Loyola's Stritch School of Medicine and has been in practice for 33 years. He is one of two doctors at Loyola University Medical Center and one of eight at Sarah Bush Lincoln Health Center who specialize in Neonatal-Perina-

tal Medicine.

Grant's time with us was short. Only four days. We held him, loved him, and certainly have never forgotten him. His birth was an event that changed Linda and me. Years later, it prompted our older daughter Kristi to become a neonatal intensive care nurse. And as I have shared often with friends, these folks are special people, truly *caring* for the patients as well as their loved ones, making the most difficult moments a bit less so, just the way Dr. Jon Muraskas proved to us is possible.

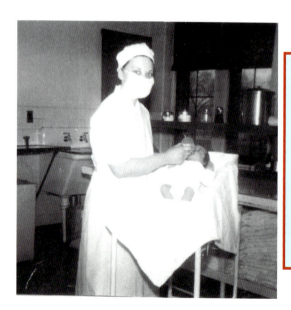

I have a feeling that Mom, on the job at the Chicago Lying-In Hospital before she returned to the farm to marry Dad, would be proud beyond words that her granddaughter Kristi chose to become a neonatal ICU nurse.

Before the Farmall "Super H" left Southern Indiana, where it had spent its first 42 years, the Armstrong girls, Kristi and Lisa, took a ride in the field.

Right, sets of the "Max Armstrong Family Farmalls" were coming down the assembly line at Scale Models, Dyersville, Iowa. It was 1998. This marked the first time that two 1/16 scale tractors were boxed together. Several 1/16 boxed sets followed from the Ertl toy line.

FIFTY YEARS AGO, ALLIED RADIO in Chicago made it possible for a grade school kid to build his own little AM radio transmitter and get on the air.

All I needed was a screwdriver, a pair of pliers, wire cutters and a soldering iron and WMAX was on the air. No license was needed from the Federal Communications Commission since the power from this unit was just 100 milliwatts. (Mom's toaster had more, I think.) And the antenna was supposed to be limited to 10 feet of wire, but mine was about 30 feet, running from my upstairs bedroom to a pole behind the old chicken house.

I could almost get a signal from the farmhouse into town, and while no one ever heard me, at least I knew it was possible they could. Pretty pathetic, huh? With my radio station, Little League baseball and those Farmall tractors, I had all I needed at age 11.

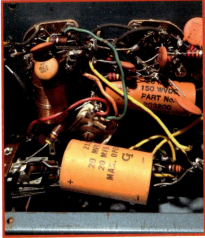

REMEMBER THE BUMBLEBEE sweaters that were big 30 years ago? My guess is that they were hot items in Purdue bookstores. And I actually thought it looked good? We were outside of Ross-Ade Stadium before a football game, and with me was the late Jerry Collins — Jerry Clough — for whom I had worked at WASK Radio, a great AM-FM station in Lafayette. He and Mike Piggott tolerated me as a part of their team all four years I was at Purdue. Both were superb broadcasters. I felt lucky to hang out with them and still am grateful for their guidance. But why didn't they tell me to not wear that sweater ... ?

— Chapter Five —

Sign On to Sign Off

t 16, I got both my driver's license and my Federal Communications Commission license, and I started working at a local radio station in the winter when Dad didn't need me in the field. I had to have the driver's license because the station was about 12 miles away across the Wabash River in Southern Illinois.

WVMC at Mt. Carmel, Illinois, was a small, locally-owned, 500-watt, daytime-only AM station at 1360 that signed on at sunrise with "The Star-Spangled Banner" and signed off at sunset with a Jim Nabors rendition of "The Lord's Prayer." That was back in an era when grade school kids both pledged and prayed to start the day.

When Jim sang, "Amen," I hit the plunger which turned off the transmitter. A second later, WSAI in Cincinnati, a 5,000-watt station on the same frequency, came blasting through the speaker. WVMC's signal was so weak that there were times when you could see the tower, but couldn't hear the station.

Working at the stations on Sunday mornings was a great experience. I loved running the station and having the place to myself on the weekends. My duties included making sure the toilet paper was in place and burning the trash out by the cornfield. But the job caused a change in my routine. I had never missed a day of Sunday School until I got into radio, but I still got plenty of religion. Three preachers would come out each Sunday morning to do church shows, and others were on tape.

The station owners bought an FM station in town and moved it out to the AM station studios where I was working. Until they automated the FM, they had an old rock 'n' roll DJ named Larry come in on Sunday mornings to keep it on the air. Well, no one had given Larry the heads up that there were preachers in the building on Sunday mornings. One of those mornings, I was in the control room at the AM station and Reverend Theron Arnold came in to do the Baptist Hour. As he prepared for his show, he was back behind the record rack, out of sight, looking through the Christian albums; our collection included such greats as George Beverly Shea, Tennessee Ernie Ford, and Anita Bryant. On the air was a reel-to-reel tape featuring another minister who was preaching about the evils of fornication.

Right on cue, Larry came bouncing through the door yelling, "Fornication! Fornication! Everybody needs a little fornication!" Reverend Arnold's head popped up over the record collection, and I said, "Larry, please meet Reverend Theron Arnold from the First Baptist Church." Larry's jaw nearly hit the floor.

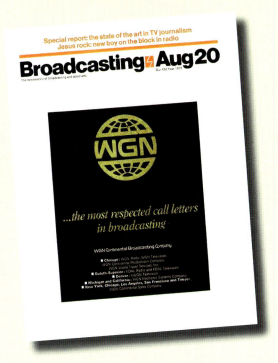

IN OUR MAILBOX ON THE FARM when I was 16 we would get *Indiana Prairie Farmer, Hoosier Farmer* and *Successful Farming.* Oh, and mailman Charlie Braselton also delivered my subscription (paid for by my field work) to *Broadcasting Magazine.* (Yeah, I was hooked at a young age. I really did know what I wanted to do when I was seven.)

I remember vividly the ad cover of *Broadcasting Magazine* from August 1973:

"WGN ... the most respected call letters in broadcasting"

I always liked the sound of that.

Historians have told me that the windows of the WGN transmitter building *BELOW* were filled in with bricks and security was tightened on the advice of Chicago police during the late '60s riots, even though this building was many miles from the protesters clashing with police in Grant Park and elsewhere in the city.

— Chapter Six —

The Blowtorch

*F*ar away from the Magnificent Mile where the WGN Radio studios are located, the 50,000-watt transmitter that powers the legendary signal to over 38 states (at night and when the atmospheric conditions are just right) is housed in a brick building on Rohlwing Road. It's south of Woodfield Mall in the northwest suburbs, and once considered part of Roselle, it was eventually surrounded by Elk Grove Village.

The WGN Radio main transmitter sends a 50,000-watt signal up a 760-foot tower, non-directionally beaming the Class 1-A "clear channel" of 720 khz. Many talented, brilliant engineers have maintained the transmitters and kept the tower energized, the signal clean, the mike open and the VU meters out of the red as we did whatever it was we did, including recently retired chief engineer Jim Carollo, who was with WGN for 40 years. The place is as neat as a pin, thanks partly to retired WGN engineer Don Albert, who manicures the lawn and chases out the deer, coyotes and foxes as he mows around the main and auxilliary towers.

One of my favorites was Ed Wilk, one of the nicest and smartest men you'd ever want to meet. Ed passed away shortly after he retired. He kept us on the air for over 30 years, and had forgotten more than most people know about broadcast transmitters and antennas. If you asked him a technical question, you would get a technical answer, complete with a sketch. He was a wonderful fount of knowledge and engineering colleagues marveled at his brilliance. As longtime fellow WGN engineer Aubrey Mumpower put it, "Ed was one in a million. His layouts, designs, and custom-made hardware are truly works of art. His kindness and generosity with his time was unrivaled for anyone with his ability. Ed had so many design ideas that we would ooh and aah over. Some of them are now in use

Ed Wilk stands next to his "babies," two of the three WGN transmitters. The main transmitter (50,000 watts) is on Ed's immediate right. The first auxilliary transmitter (50,000 watts) is to Ed's left. The second auxilliary (10,000 watts) is in a separate building.

across the country. I was proud to call him a friend."

Former WGN general manager Tom Langmyer said, "Ed was a very positive and kind person, a man of great character and integrity."

I couldn't have said it better. I thought a lot of Ed.

Dad's log chain. Every tractor we had when I was a boy carried one of these. They were used often, to pull another tractor from the mud in a wet spring planting season, or to get a motorist's car out of the ditch on a snowy winter night.

And when I headed to Chicago to start my job at WGN, Dad made sure I had this chain in my car, wrapped up in an old burlap feed bag ... just in case.

This photo is from the mid-'90s at WGN Radio's "PumpkinFest" at Jim and Esther Goebbert's farm in Northwest Cook County. The guy in the bibs was being "interviewed" by both the legendary Bob Collins (notice the six-shooter strapped to his hip) and Orion Samuelson. Behind Bob, was consulting horticulturist Jim Fizzell, and on the far right was our long-time superb assistant and great friend, Lottie Kearns.

— *Chapter Seven* —

Bob Collins

*U*ncle Bobby picked on me, like he did most everybody else. And it was wonderful, because as he picked on you, the listeners sensed that maybe Bob liked you, and that endeared you to the listener as well. Bob joined the station in 1974 and I came along in the fall of 1977. He had a tremendous impact on me, as he did with everyone he worked with.

Bob Collins was a tremendous natural talent. He could be clowning around one moment, and the next moment be handling a very serious issue. He was well-read and abreast of current events, and was connected to a lot of people. What a lot of people didn't realize was that Bob was shy. Here was a guy who was so outgoing on the radio, but at appearances, he could be kind of shy.

Before Bob moved to mornings to take over for Wally Phillips, he was on WGN Radio in the afternoons. I would sit in the studio, four feet to his left, reading the market prices and agricultural news. I took the information seriously. It was my career, it was my livelihood! But Bobby did everything he could to distract me and make me crack up.

Bobby used a variety of methods. He would reach over with his cigarette lighter and set my script on fire. One day, he was bouncing quarters off my microphone. Other times, he would pull out what he called a medical journal, which was actually an adult magazine, and inched it up onto my script as I was trying to read the prices. And it was opened to one of the "important" pages. He was forever doing things like that. I was so glad when Tom Skilling came on because Bobby started picking on Tom and was a little easier on me.

The off-air conversations with him were so much fun. I'd set him up a little bit by dropping a line about 27 seconds into a 30-second spot. I knew what the result would be. The mic would come on and Collins would be cackling. He'd often say, "Oh man, if that relay on the microphone ever stuck, this place would be a K-Mart parking lot." We grew up hundreds of miles away from each other and took very different paths. But I remember sitting with him in the studio off the air during commercials or news, and he would start some profane joke, and I could supply the punch-line. Like something you heard when you were seven or eight years old. And it might've been something I hadn't heard in 30 years. And yet Bob would lay down the expression and I was able to finish it. How did that happen? What was the connection? How was it that I was able to know that same joke?

As Orion has often shared, he and Bob were not friends at first. But it didn't take long for that wonderful friendship to blossom, and Bob welcomed me into the fold, too. And that was a wonderful thing for me.

Before he was hired at WGN, Bob was working in Milwauke. I recall hearing the tape of when he came to Studio B at WGN Radio on Bradley Place and did about an hour audition. It was interesting to hear him introducing songs, reading a lot of live copy. He had a unique style, was very bright along with being naturally funny, and, obviously, became a great addition to WGN.

It was a tremendous era at the station with Wally doing the mornings and Roy Leonard in the mid-day with a split shift around the *Noon Show*, which Orion and I did. Bob also had a split shift: two to four, and seven to nine. And during that period of time between four to seven, he quite often wound up at Schulien's on Irving Park Road. As an impressionable young man, just arrived in Chicago, I went with him a couple of times and decided that I wasn't up to that challenge on a regular basis.

Uncle Bobby also constantly traded cars. During the period of time that I knew him, 1977 to 2000, I would say he had 60 different automobiles. That may be a conservative number. During the energy crisis, when the lines were long at the service stations, everybody was looking for fuel-efficient automobiles. Bob bought a VW Rabbit with a diesel engine, 42 miles to a gallon. And after he'd put about 5000 miles on it, he said, "Hey Bub (he called me Bub), you want to buy that Rabbit from me?" I said, "Yeah, I'd love to." I bought it, and I recall pulling into the truck stops, the only places you could buy diesel then, and how the truckers reacted to my yellow car. It was school bus yellow, banana yellow. The ugliest car I've ever owned. But the big problem with the car was that the heater wouldn't work. And if you remember, the winter in 1980-81 was one of the coldest on record. I'd drive into Chicago wrapped in blankets. I took it to three dealers and no one could fix that heater.

Well, after I told Uncle Bobby about the problem, he would cackle on the air, "Hey Bub, how's that yellow lemon I sold you?" He took great delight in my struggles with that car. The one he had sold me!

In the summer of 2014, I took a picture of a yellow diesel VW Rabbit I spotted in a parking lot in North Carolina. Just like mine. It brought back a lot of memories. Bob also teased me about a tractor I bought from my dad, the Farmall "Super H," the one I learned to drive on.

He accused me of stealing from my elderly father.

Ironically, my last conversation with Bob was about airplane safety. It was February 8, 2000, and it was during Tom Peterson's eight o'clock news. I was sitting in the studio with Bob, and I had flown to Western Illinois University that weekend on Air Orion with one of Orion's pilots. An experienced and avid pilot himself, Bob asked, "Who flew you?" And as I told him, he said, "I don't know him. Is he a good pilot?" And I said, "Oh yeah, he's very good and I felt very comfortable flying with him," which was the case with both of Orion's pilots, Phill Wolfe and Jerry Lagerloef. And that was it, the last conversation we had. At about three that afternoon, Bob died after a mid-air collision over north suburban Zion.

All these years later, Bob Collins continues to be remembered. Farmers come up to me, 200 miles from Chicago, and get teared up talking about him. He had so much impact on them as they sat in their combine and tractor cabs, listening as they worked the fields.

Bob's funeral was the following Saturday at Holy Name Cathedral. I worked through the night in my home studio to record The Noon Show so we could be there early for the service. For the broadcast I used outtakes from Bob's audition tape, recorded when he came to apply for the job 26 years earlier. I still have his audition reel, by the way.

As Linda, the late Dick Sutliff and I got into the taxi cab in front of Tribune Tower for the trip to Holy Name, the cabbie actually had WGN Radio playing in his car. So the first thing we heard after closing the taxi door was my voice and then Bob's. The cabbie not only did not turn the radio down so we could talk in the backseat, he turned the radio UP, way up, after saying, "Thank you, and I am sorry for your loss."

Bobby's funeral was an overflow crowd with all types of people paying their respects. Lined up with his WGN family were the most powerful politicians and the wealthiest people in the state, right along with a homeless woman weeping as she listened to her transistor radio. She was just as affected by Bob as anyone else. He was just a tremendous guy.

Uncle Bobby and "Bub"

Robert Lind

On a rainy Saturday in 1995, Bob Collins stood on my Farmall "Super H" drawbar. We were at another "PumpkinFest," at the Goebbert Pumpkin Farm. I had just acquired the "Super H" at Mom and Dad's auction and brought it from the farm in Indiana that spring. Uncle Bobby told his vast radio audience, "This is the tractor Max stole from his poor old mom and dad."

A print ad in 1983 urged people to listen to Orion and me. They didn't have to try too hard to hear us. At one time, we were doing 190 broadcast inserts a week on WGN and the network!

On the This Week in AgriBusiness *set at our Aurora studios with Orion*

— *Chapter Eight* —

"Big O"

*I*t may surprise you to read that I never intended to be an agricultural broadcaster. As a kid on the farm, I listened to the big Chicago AM stations that boomed in from 300 miles away. I loved hearing icons like Clark Weber, Dick Biondi and Larry Lujack, and was thrilled to later get to work with some of those guys, like John Landecker, Joel Sebastian and Lyle Dean. And while I dreamt of being on the radio in Chicago, I really wanted to be a newsman.

Although I never imagined a farm broadcasting career, I do remember seeing and hearing Orion when I was in college. He was doing a remote broadcast from the Purdue Union Building for the Indiana Soil and Water Conservation Districts. I was intrigued, but felt the specialty of farm broadcaster was way too narrow for me. I didn't think I would ever have a chance to work in a major market like Chicago.

In hindsight, of course, it would have been a terrible mistake for me to have devoted my energies to chasing police cars, ambulances and politicians, rather than trying to corral steers, heifers, barrows and gilts, and the people who raised them. Little did I know I would spend these four decades as Orion's colleague and business partner.

I met him in 1977 on the floor of the Chicago Board of Trade. I was working for the Illinois Farm Bureau at the time, only two years out of Purdue. After trading had finished for the day, the CBOT put on these grand dinners for the press, right there in the pit where a short time earlier, traders were jammed elbow-to-elbow, yelling at the top of their lungs. I told Orion that I'd love to have a chance to talk to him if he ever planned to expand his staff. It wasn't too many weeks later that my predecessor, Bill Mason, left WGN. That left an opening and Orion called me.

When I went to work at WGN, I was the youngest on the radio there by about 8 years. I thought I'd last six months, but if it was only six weeks I figured having WGN on my resume might look good.

It's always been a thrill to work with Orion. I tell the joke about people confusing us every now and then. Somebody would come up to me and say, "Orion, how are you doing?" One time my wife, Linda, and I were in line at a theater in Bloomington, Illinois. A woman came over to me and said, "I would recognize you anywhere! I listen to you all the time and watch you on television. I think you do a fantastic job, Orion!" I responded, "Well, thank you very much, and I'll tell Max you said hello, too!"

Something I've admired about Orion — and, frankly, marveled at — is his tremendous work ethic, which has not wavered as he's gotten older. Here's a guy, 81 years old, driving 55 miles to the city at three in the morning, four days a week. The fifth day, usually Wednesday or Thursday, he does his market reports for *The Steve Cochran Show* from our TV studio in Aurora. I've had to run hard just to stay up with him. Only in the past few months have WGN management and Orion's wife Gloria been able to convince him to cut back, but only a little.

Orion's tremendous to work with, and is generous with his spirit of cooperation. I readily admit how much I've benefited from my association with Orion, and from the rest of the WGN family. I've been so blessed just being there with those folks. I was there full-time from 1977-2009, and I'm still on the air on *The Saturday Morning Show* with Orion, often broadcasting from my home studio in North Carolina. It's a hoot to do the show as the sun is coming up over the Carolina pines, while it's still dark in Chicago.

There has been no better friend to the American farmer than Orion. His grasp of the ag issues and his eagerness to support producers is unrivaled. And he's never been afraid to give his opinions, even though they often upset many in the audience. I've shared in his dedication to the farmer and a love and appreciation for the agricultural community. We both have the desire to tell their story to others. Orion pioneered this and it was easy for me to ride along with him. And it's gratifying for us now to see how excited others are to tell the story. So many farmers in their 20s and 30s are on social media spreading the word about the great things happening in agriculture. They're blogging, too. They do a good job of cutting through incorrect information and misunderstandings about how food is produced. But long before social media and the Internet, Orion and I were telling agriculture's story on radio and television, and at public events.

Frankly, it is a challenge to stay on top of the issues. It's a complex industry and when it comes to farm policy, there are no easy answers. It's a challenge to explain the history and why things are done as they are, how and why we've come to this point in agriculture policy and trade, and the intricacies of crop production and transportation that allows food to get to the consumer.

Thankfully, we always got a lot of opportunity through the support of Bob Collins, Wally Phillips and Roy Leonard with their huge audiences.

They fostered that appreciation for agriculture and always let us interrupt their conversations with a parade of stars and talent for our market reports. Many times, they would follow-up with questions about agriculture. Bob Collins would call back to our office, on the air, and ask, "Where's the cow editor?" I'd respond, "I think you've got him, pal."

Live Broadcast

WGN Radio 720

Noon Show

featuring
Orion Samuelson
and
Max Armstrong

WGN Radio 720 AM

For decades, **The Noon Show** *was a hugely popular show on WGN, often a "must listen" during the break for dinner (what we now refer to as lunch) when farmers across the Midwest came in from the fields. Orion and I often took the show on the road, as shown above.*

ORION OFTEN MENTIONS his airplane, Air Orion, during his broadcasts, and many of the controllers in the region recognized the tail number. One morning, the air traffic controller recognized the call sign for Orion's plane: "Uh, 9999C Charlie, where are you taking the Big O today?" Captain Phill Wolfe glanced over at me as we flew at about 4,500 feet, pushed the transmit button on the yoke, and said, "Uh, I got the other guy today." In a split second, the radio chirped again. "Oh, you're hauling 'Tractor Boy!'"

WGN

I guess sideburns were "in" when this WGN
Radio publicity piece was put together in 1978.

Chapter Nine

Wally Phillips

O ne beautiful spring morning on the northwest side of Chicago, I was sitting in my car at a busy intersection. It was about eight o'clock. Everybody had their windows down and their radios on, and as I looked around, I realized people were laughing as I was laughing. I turned down my radio and I was hearing Wally Phillips's voice coming from the cars all around me. It was a reminder of what a ratings giant Wally was. No other radio personality was even close to him in the market.

Wally came from Cincinnati and was hired by Ward Quaal, who was responsible for bringing in other incredible talents like Bob Bell — "Bozo" — and Frazier Thomas. And Ward brought in Orion, too!

Wally was first to start taking telephone calls on the radio. And he would originate calls, too, often to the amusement of his audience at the expense of the person he called. Wally is why the FCC came up with the regulation that you needed to tell people that they were on the air, or that you were recording them to play back on the air. Wally had a terrific producer, Marilyn Miller, who skillfully orchestrated callers, initiated calls for Wally and acted as a gatekeeper for him. We used to sometimes laugh about it and call her "Dragon Lady" because of how aggressively she protected him. But she had to erect barriers because there were so many people who wanted to get to Wally, and be on the show. She was there to sort through the requests and make sure his show was the best it could be.

Orion and I had our own version of the "Dragon Lady." We were so blessed to have Lottie Kearns with us for over 25 years. She didn't grow up on a farm and knew nothing about agriculture, other than what she picked up sitting 10 feet away from us. It was amazing how much she learned about agriculture in those years. Lottie had unbelievable energy, a great work ethic, and was indispensable to Orion and me, arranging all our travel for speaking engagements. She also produced our shows. When farmers called from their combines during harvest, Lottie was the first one they spoke with.

This was Wally Phillips before he started buying his hair, joking with Orion and my predecessor, Bill Mason

WGN

— *Chapter Ten* —

Roy Leonard

A tremendous professional. There's no other way to describe Roy. He was so kind to his guests, and all of the stars visiting the studio felt so welcome. They would tell him things they probably wouldn't tell others, simply because they liked him and were so comfortable with him. You could see it on their faces.

When Orion and I were doing our market reports, we'd sit six feet away from Roy and his guests. It was a thrill, and Roy would always include us. I'll never forget when John Denver was in the studio and Roy introduced us. He reached out to shake my hand and introduced me to Annie, his wife at the time. This was when WGN Radio and TV were together on Bradley Place in northwest Chicago. It was a constant parade of stars, politicians and characters of all sorts in that studio, partly beacause Phil Donohue's show was being done down the hall each day. You never knew who you'd run into. I met Billy Carter, the late brother of President Jimmy Carter, in the restroom.

So many of us admired Roy because he was not only connected with famous people, but he was very family-oriented. Every year at Christmas time, the whole family, Sheila and the six boys, whose names all began with "K," were in the studio with him. And he would talk about Sheila all the time. She was such an important part of his life, and he of hers. They were both so genuine. There was not a phony bone in Roy Leonard's body. That's the thing I really admired about him. In an industry with big egos and people wrapped up in themselves, here was this guy who was larger than life in the radio industry. And you'd never know it.

To a fault, I suppose, I struggle with people who aren't humble. I guess it's something I learned from my parents. No matter what you've accomplished, what you've done, or where you've been, and what you've amassed, at the end of the day we're all in this together, all in the same boat. There's nothing I know of that justifies treating someone as if you are high and mighty. Roy had every right to be one of "those" people, but he wasn't. He was a genuinely nice, down-to-earth person, a huge talent, and it was an honor to be his colleague.

Roy Leonard with Bob Collins at WGN Radio, circa late '90s

Chapter Eleven

I are a

Professional Broadcaster

O ne day in my early years at WGN, when only a few hundred thousand people were listening, I was reading a live advertisement for the Mill Run Theater at the Golf Mill Shopping Center. I confidently read the copy and finished it off with, "Be sure to stop in to see Steve and Eddie!"

At the moment I said that, out of the corner of my eye, I could see producer John "Mad Dog" Madormo jump out of his chair. Broadcast engineer John Bobera Jr. looked up through the control room window with a puzzled look. Staff musician-record turner Dom Geraci smiled ear-to-ear from behind the turntable desk. Mad Dog hit the talk back switch from the booth and said in my ear, "You said Eddie. Eddie! Did you perhaps mean to say Eydie? Steve and EEEEdee?"

"Yes, of course," I corrected myself, "Steve and Eydie will be appearing at the Mill Run Theater!"

Dom, a trumpet player by profession, said, "Just keep going, Kid. Don't look back."

A statue of Jack Brickhouse sits prominently on Michigan Avenue just outside the WGN studios.

— *Chapter Twelve* —

Jack Brickhouse

When I think about legendary, long-time Chicago sportscaster Jack Brickhouse, I can't help but remember what a class act he was.

For 41 years, out of the nearly nine decades that the Chicago Cubs games were on WGN Radio, Brickhouse was in the booth. Unlike many other sportscasters, "Brick" was able to do far more than sports. Yes, he did the Cubs, the White Sox, the Bears and pro wrestling, but he also was on the floor of the political party conventions with a microphone.

In the late '80s, I was standing at Ohio Street waiting to walk south along Michigan Avenue toward the Tribune Tower. Somebody put their hand on my shoulder, and I turned to see, looming over me, Jack Brickhouse. He said, "Kid, are you headed back to the Tower? I'll walk with you."

Walking down a busy Chicago street with Jack Brickhouse was a spectacle! Cab drivers, bellmen, businessmen ... they were drawn to him like a magnet. It was SO fun to watch. And he was SO gracious.

This was the board located offstage that was used to keep track of the show order for Ray Rayner and His Friends *on WGN-TV.*

— Chapter Thirteen —

Ray Rayner

*I*f you grew up in the Chicago area, the picture I took on the previous page might bring back some warm, fuzzy memories. It was the in-studio format board for Ray Rayner's show at WGN-TV. *Ray Rayner and His Friends* followed *Top O' the Morning* with Orion, a show I wound up hosting quite often in the late '70s and early '80s when Orion was at a farm meeting in Teutopolis, or Towanda, or Tel Aviv.

Ray was a great, genuine talent and was fun to be around. He was a terrific entertainer, a talent he realized he had as a prisoner of war during World War II. He was the navigator of a B-17 that was shot down over France in 1943. During two and a half years as a POW, he entertained his fellow prisoners and his German captors. At WGN, the "Friends" in his show's title included his stagehands. Dave Abrams ("Chauncey"), Mike Cusick and Johnny Dial actively participated in the show, would throw out lines, talking back to Ray from off the set.

Ray handled the Illinois Lottery drawings in the first few years when it was just a once-a-week event. In 1979, Ray was on vacation, and no one had been scheduled to host the drawing. The program director of WGN-TV came to me and said, "Kid, do you want to do the lottery tomorrow?" Uh, yeah, sure! I had never even played the lottery. Worse yet, I had just changed the fuel pump on my Chevy Blazer the night before, and my fingers looked like I had just changed my fuel pump the night before!

I stayed up all night trying to learn the lottery workings and scrubbing my hands, getting my fingers ready for the close-up camera shot when each numbered ball came up the tube.

I was no Linda Kollmeyer.

THE STATION WAS BROADCASTING in color. I'm not sure why the photo above is black and white. The engineer behind the camera was "Deacon." I think his real name was Vern Plateau. Partly visible to the left, behind the *Top O' The Morning* set, was Ray Rayner's set. It was enjoyable being in the studio before Ray. It didn't take long for the set to be changed, which meant Chelveston the Duck would be brought out.

On my way out of the studio at 6:30, as I would pass Chelveston's cage on the floor in the hall, I would gently give it a little nudge with my shoe ... just to make sure he was ready for Ray. Now it can be told.

Shown above are Cookie (Roy Brown), Bozo (Bob Bell) and Frazier Thomas. They were superb companions for kids. For many years it was LIVE television during the noon hour. Frazier would arrive at the station in the office next to mine around 6:00 a.m. to start writing the script for the day. I cherish the memory of those visits.

Tickets for Bozo were the hottest thing in town, nearly impossible to get, and kids sometimes waited years to get on the show. One morning, I was stopped for speeding en route to work on the Kennedy Expressway. (66.6 mph, the officer said.) After asking where I was going at four in the morning, the officer asked if I knew Bozo, and then explained that his kids had four more years to wait for their tickets. Long story short, between him and his partner in the squad car, they had 10 kids. 66.6 cost me 12 Bozo tickets! It was a good thing I was on good terms with Mae, the ticket lady.

This is the opposite side of the postcard sent by Bozo's Circus. *I guess they thought the card would pacify the kids and parents while they waited more than eight years (that's right!) for tickets.*

Chapter Fourteen

Tom Skilling

*T*alk about big shoes to fill! In the late '70s and early '80s, I was a third-string, part-time fill-in as the TV weatherman on the rare days when both Tom Skilling and Roger Triemstra were gone. I wound up doing a few nights a year for several years, and I'm sure that every minute I was on the air, viewers were wondering, "Where's Tom? Who is this hayseed?"

Tommy was just getting into the electronic weather graphics at that time, and every time I would substitute, I had to learn the computer program anew. "You'll have no problem with it, Max," Tom would say, and he would leave a few scribbled notes on a yellow legal pad for me to decipher. Truly, it would take me four hours to prepare for three minutes on the air!

Tommy came to WGN from a TV station in Milwaukee where he had been doing the weather with a puppet, but not by choice. Tom says Albert the Alleycat was a creation of WITI-TV's management team and had been at the station for years before Tom was hired. Albert had been paired with Ward Allen on the weather programs on WITI for more than a decade and Tom's hiring was part of an effort to phase Albert out — a process which halted when the station was literally swamped by viewer mail protesting the move.

Tom is a good guy and "the real deal." Tom IS the franchise for Channel 9's news and is a truly dedicated weather scientist.

No media weather forecaster — heck, no media personality, period — has more credibility than Tom Skilling.

My old buddy Tom Skilling as we visited in his office many years ago when it was crammed in behind the set at WGN-TV.

I'm not sure why I have the "hand-caught-in-the-cookie-jar" look, but this was just after we moved into our shiny new studios at the Trib Tower. For my younger readers, those round, flat things in the foreground are turntables. Once upon a time, that's how we played music.

FIRST PICTURES OF STORM DISASTER
HERALD CHICAGO EXAMINER

Telephone Main 5000   FRIDAY, MARCH 20, 1925.   C"   TWO PARTS   PRICE 5 CENTS

1,000 DEAD, 3,000 HURT LATEST TOLL OF TORNADO

In the Twinkling of an Eye, Murphysboro Was No More

The horrific Tri-State Tornado of March 18, 1925 blasted parts of three states. It was the deadliest tornado in U.S. history, killing nearly 700 people who never had a chance. (The Chicago Herald Examiner *had the toll a little high.*)

— Chapter Fifteen —

Tri-State Tornado

My dad was eight years old when the deadliest tornado in American history hit. The Tri-State Tornado started in Missouri early in the afternoon of March 18, 1925. It went across southern Illinois, crossed over the Wabash River and into southern Indiana. It went a distance of 200 miles over three hours and killed 695 people. Among its measurements: the trail of devastation was the longest recorded in the world. In addition to the 695 deaths, over 2,000 were injured, and some 15,000 homes were destroyed.

Radios weren't common yet in rural America, and when a tornado hit a town, it usually ripped out the telegraph lines. So there was no way to send word ahead, and each town was hit with no warning. My dad's family lived in the country, many miles outside of Owensville, Indiana. He was on his way home from school in a horse-drawn wagon with other kids. My dad remembered seeing his lunch box flying down the road as the horse spooked and ran. The wagon was turned over, the horse went down and the children were thrown into a ditch. It left quite an impression on him. He saw badly injured and dead horses as he and the other kids tried to find their way home to their farmhouses.

*After being notified of the 1925 tornado
tragedy, International Harvester in Chicago
loaded trains with McCormick-Deering 10-20's
and shipped them immediately, saving the
crop season for farmers. These photos show
the scene near my hometown of Owensville.*
(Thanks to Stanley Douglas for the photos.)

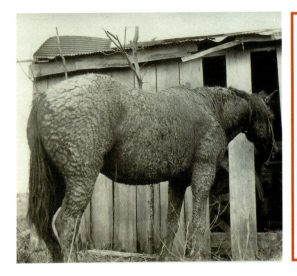

From the Armstrong Family archives comes this picture taken in the wake of the 1925 Tri-State Tornado.
Mud-covered Old Paint appears to be looking into the damaged shed to see if the tractor got roughed up as much as he did!

Dad always said he never saw a funnel cloud. "It was just as black as night," he told us. I thought maybe his memory had been fogged. He must have seen a funnel. But Tom Skilling told me that the path of the tornado was so wide that Dad, right in the middle of it, likely never saw the characteristic funnel-shaped cloud.

And it was fast! Records show that at the point where Dad experienced the storm, it was moving at 73 miles per hour.

The timing was terrible because farmers were just getting ready to plant. Many of the horses had been killed or injured and machinery had been wiped out. There was no way to plant the crops. A Farm Bureau representative called International Harvester in Chicago and asked for help. Within hours, IH had a trainload of 20 tractors and plows on the way to Indiana. Farm Bureau called IH, thanked them, and said there was a similar problem in another town. And IH immediately sent another 12 tractors.

Forty years later when I was growing up, the big tornado was still evident. Here and there you'd see a piece of tin or other debris sticking out of a tree trunk. The tree had grown around it.

Most of the tornado's survivors are long gone, including my dad, but I'm so thankful he survived that deadly tornado!

WE ARE BLESSED TO HAVE great friends who have crammed a lot into their lives. World War II veteran Harold Steele of Princeton, Illinois, is one of those.

For 13 years, Harold was president of the Illinois Farm Bureau (my boss for two-and-a-half of those), and he served four years as chairman of the federal Farm Credit Administration in Washington, D.C., appointed by President George H.W. Bush. Pictured above with me in May 2014 at the age of 92, Harold was "as sharp as a tack."

I am so thankful I didn't miss Harold's on-farm Armed Forces Day tribute. A social media friend who was there called Harold "the greatest of The Greatest Generation."

Amen.

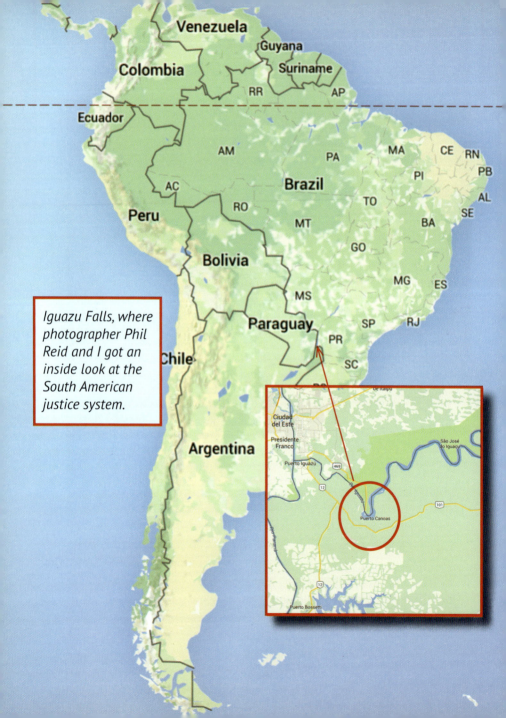

Iguazu Falls, where photographer Phil Reid and I got an inside look at the South American justice system.

Chapter Sixteen

The Camera Thief

O rion and I have had the pleasure of working with a great crew of video production professionals on our weekly TV show, *This Week in AgriBusiness*, which we produce each week at a studio in Aurora, Illinois. Phil Reid has the longest tenure, going back 40 years, and over the years, Angelo Lazarra, Ryan Ruh, Marilyn Reid and Kristin Decker joined the team. We've traveled the world covering issues important to agriculture, and one year, Phil and I were in Brazil near the border with Paraguay. We were traveling with some farmers, and a tour bus took us to a huge waterfall — larger than Niagara Falls — called Iguazu Falls. As we got out of the bus, we were assured by our guide that it was fine to leave Phil's video camera on the bus with the driver.

After seeing the spectacular falls, we went back to the bus and found that the driver had fallen asleep, and someone broke into the back of the bus and stole our camera. It was gone!

We had been warned about pickpockets. This was a very busy and congested area and we had neon signs over us saying we were Americans and suckers for the pickpockets. You could feel a hand go into your pocket now and then. Losing a few bucks was one thing, but a television camera in those days was a $50,000 investment. We thought it was gone forever.

We went to our guide and told him we wanted to offer a reward. A police officer put out a dispatch that there was a reward for a stolen camera. It was found in 20 minutes! They caught an 11-year old kid getting into a taxi cab with it.

Our challenge was to then get it back from the police. This became a significant community event. Phil and I went to the police station to fill out the paperwork. In a dimly-lit room, there was a guy sitting at a manual typewriter with carbon paper. We could tell this was going to be a lengthy evening. They showed us the kid in a jail cell. The cells, dank and lit by dim, solitary lightbulbs hanging from the ceiling, were packed with kids. We could only see eyes.

Meanwhile, the cops had called their wives and asked them to bring pizza and beer for a party. So Phil and I enjoyed a party at the police department. They even called the local newspaper and a reporter came over and took a picture of Phil holding the camera and the police holding their firearms.

Not his best side, but here's an action shot of Phil Reid, the longtime TV photographer and producer who does his best to make Orion and me look good.

Max Armstrong

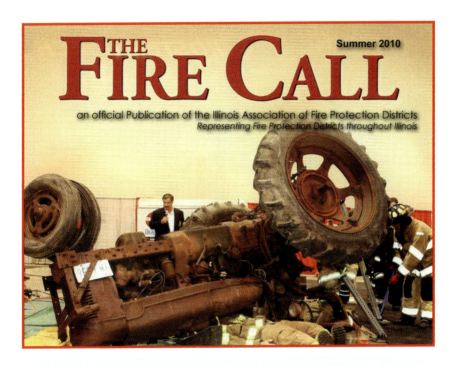

Summer 2010

THE FIRE CALL

an official Publication of the Illinois Association of Fire Protection Districts
Representing Fire Protection Districts throughout Illinois

ON THE COVER OF the September 2010 *The Fire Call*, three firefighters demonstrate how to save a life from a tractor roll-over accident, a leading cause of farm deaths.

The demonstration, sponsored by the Illinois Fire Store, was held at the annual meeting of the Illinois Association of Fire Protection Districts in Peoria.

One "victim," a dummy, is visible under the tractor. A second dummy can be seen with a microphone on the other side of the Farmall "H."

The only things left from a disastrous combine fire on the Armstrong farm

Chapter Seventeen

The Combine Fire

A combine fire is an event you never forget. In my case, it was in 1967. Our two-year-old — nearly new by farm implement standards — International 203 combine caught fire that afternoon while I was at football practice after school. I was 14, too young to fully understand the ramifications of such a new and important piece of equipment being destroyed, but I could sure see it on Dad's face that night. He hadn't been driving the combine. Uncle Frank had, and he had nothing with which to fight the fire. I have never forgotten that crisp, partly cloudy October evening.

To help remind me, I guess, of what the day was like, I saved the switches shown on the opposite page. Salvaged from the burned carcass of that 203, they still have a little International Harvester red paint on them. I pull them out and play with them now and then, bringing back some unforgettable harvest memories of Dad and Uncle Frank.

I'm not sure when and why my fascination with firefighters began, but I'm amazed at people who run into a building when others run out. Even in the most difficult situations, they rise to the occasion. This is what they train for, and they not only risk their own lives, their service is not without ramifications in their personal lives. There needs to be more attention to that.

For several years, I was the MC for the Illinois Fire Marshall Awards in Springfield, an event honoring firefighters who died in the line of duty and others who performed acts of heroism. It was my role to take the official reports, which terribly understated the drama in the description of the event: "Engine 53, Ladder 25, Battalion Chief arrived on the scene to find the building fully involved." Period. It would take me four or five nights to write the script so the incident would come to life in the two to three minutes I was allotted to explain what happened. What was the temperature that night as the firefighter was hanging on one arm pulling a child out of a fiery window? Was it 22 below?

There are usually 2,000 people there at the ceremony. For every firefighter who has died in the line of duty, they'll have a truck from his department sitting there. There'll be five or six trucks around the stage. It was always a somber event as we remembered those who had given their lives, but often, the hero being honored was reunited with the person he or she saved. Yes, emotions ran high and tears flowed freely. It was always a remarkable event and I was honored to be invited to take part.

Sitting in front of me each year were the union firefighters from the state's biggest cities, and alongside them were the volunteers from the smallest downstate hamlets, the people whose regular job might be running a grain elevator in town. It was a reminder that the flames are just as hot in Browns, or Bone Gap, or Bushnell as they are on Belden Avenue in Chicago.

Few things in my adult life have been as satisfying and enjoyable as my 22-year appointment to the Board of Fire Commissioners for the Lisle-Woodridge Fire Protection District, which is about 25 miles west of Chicago's Loop. I was privileged to play a minor role in this fine agency, hiring and promoting these outstanding public servants, full-time firefighters and paramedics.

These are the hands you'd want reaching into the car to help your family, and the feet you'd want coming up the sidewalk in your time of need. The men and women I have come in contact with here have made me a better person, for sure. I am so thankful for them and pray for their safety as they continue to serve.

I was a volunteer firefighter for four months in the tiny town of Towanda, in McLean County. It was just before I took the job at WGN, and Orion still teases me about my foot-dragging in taking the job because I didn't want to leave a volunteer fire department.

By the way, it may surprise you to know that 75% of America's firefighters are volunteers. In thousands of farming, ranching and mining communities they are the backbone of the town. And ... they still make house calls!

As I ended a 22-year run with the Lisle-Woodridge Fire Protection District, I was honored to be standing with this group of everyday heroes.
Left to right: *Pete Olson, Rudy Chmelik, Chief Tom Freeman and Jim Slater*

This photo, from the '70s, shows Dad combining beans with our IH 715. It was one of IH's mid-sized combines, and a good performer. This wasn't the one that burned.

CHICAGO IS LIKELY THE ONLY big city in the world where a farm combine is in a museum. It's in an exhibit called "The Farm" at the Museum of Science and Industry. This superb exhibit exists because of Sharon Covert of Tiskilwa, Illinois. She took her grandkids to the museum and found no ag exhibit. Sharon spearheaded an effort to get a display reflecting all crops and animals that are part of Midwest agriculture.

Tens of thousands climb up into the combine each year, and if you go, you'll hear a voice on the radio in the cab you may recognize. I am proud to be heard there.

I'M MISSING OUR OLD JACK RUSSELL, Toby. I guess I never realized the great gift from God that pet companionship is until he was gone. Kristi, Lisa, Linda and I were so blessed to have this guy with us for 14 years.

With sincere thanks to Cowboy Poet Baxter Black, I share a portion of his superb piece "Just A Dog."

People debate if dogs have a heaven. I'm not sure that matters. What is heaven to a dog? Enough to eat, something to chase, shade in the summer, someone to scratch your ears and pay you a little attention now and then.

All I know is you added to our life. Companion, listener, guardian and connection to a part of nature we tend to overlook because we're too busy worrying about the minutiae of life.

You reminded us to appreciate a sunny day, a bone to chew and a kind word. You'll be missed around here.

You were "just" a dog, but you'll be in my heaven, Toby. Rest in peace, old friend.

Stories from the Heartland

IT'S NO SURPRISE, I SUPPOSE, that farmer funerals often involve antique tractors and other implements. After all, farmers are known to
develop attachments to the machines they spent countless hours on,
and years later spent countless more hours restoring to their original
condition. And there's such an allegiance to color!

LEFT: In Ohio, one of
farmer Bob Wilson's
last requests was
fulfilled.
BELOW: Old Farmalls
and other equipment
escorted 82-year-
old Merlin Anderson
to his final resting
place in Elkader,
Iowa. Merlin was an
IH man through and
through. He even car-
ried an IH M1 Garand
30.06 rifle during his
service in Korea!

Danny Anderson

ANOTHER FUNERAL TRIBUTE we often see in rural America is the tractor and/or combine salute along the route between the church and the cemetery. The tribute above was near Decatur, Illinois.

In 2009, Bud Moit, a well-known farmer and John Deere collector near my hometown, passed away. His casket was hauled through Owensville on a wagon pulled by a Deere tractor, and driving it was the local John Deere dealer. Farmers are sometimes buried in a John Deere casket or an IH casket. If one isn't available, I've known guys to take the casket home, remove the handles, spray paint it, put logos on it, and get it back to the funeral home!

For Dad's funeral, the funeral director, a long-time friend, was kind enough to let me drive the hearse, and brother Steve rode "shotgun." Of all the places I have been and honors I have received, truly nothing quite compares with the privilege of taking our dad on his last trip on this Earth.

I've always said that when my time comes, Mrs. Armstrong will probably have both of my tractors buried with me.

Betsy Shiverick

PERHAPS THE MOST FAMOUS PHOTO taken by Tom Hoy was shot in the D.C. Armory in 1962. He took a risk because all of the other photographers were IN FRONT of President Kennedy.

Tom was a Cincinnati boy who spent most of his life in D.C. He started as a news photographer with the old Washington Evening Star in the 1950s and 1960s, and then spent nearly 30 years as PR director for the National Rural Electric Cooperative Association.

Tom was a superb public relations practitioner. I could never turn his story pitches down, and really didn't want to.

His storytelling was legendary, and having dinner with him was unforgettable. Tom told about ringing Dick Nixon's doorbell on a Sunday afternoon, traveling to Cuba, and being one of the few photogs at the White House in the wee hours of the morning when JFK's body was brought back. As Tom related the stories, it was not with arrogance or ego, but with pride of having had a front-row seat to some rich history.

Seeing the wall of Tom's photos in his house was like thumbing through the pages of my high school U.S. History book.

— Chapter Eighteen —

A Street to Nowhere

*I*n the town of Penfield, Illinois, I was honored by the I & I Club (the Illinois and Indiana Antique Tractor and Gas Engine Club) with my own street. It is two blocks long and, fittingly, is a dead end. It is one block from the only business in Penfield, The Last Call, a bar. There are no permanent buildings on Max Armstrong Street, but you'll notice that in this photo, taken at the 2015 Historic Farm Days Show, there is a strategically located, temporary structure.

Upon closer examination of the photo, you can see that no grass grows at the base of the leaning Max Armstrong Street sign. We are told that every male dog in town has come to "appreciate" this spot, and frequently, apparently.

One of my best friends and a fellow red tractor nut, Darius Harms

Chapter Nineteen

Mr. Harms

Other than Linda, Darius Harms is my best friend in the world. And there are probably 10,000 people who would say the same thing about him. He's one-of-a-kind! Darius is an amazing character and has touched so many people. I would guess that Darius spends more than 10% of his time visiting hospitals, nursing homes and funeral homes. He truly cares about people, and they clearly care about him.

Darius is one of the country's foremost experts on the old International Harvester Company and its equipment. He's collected a lot of stuff that's just sitting in sheds, including some rare tractors and other equipment, but he can tell you why and how each piece was developed. I've learned so much from Darius about farming and life.

My first encounter with Darius was at the Indiana State Fair where Lottie Kearns and I were producing a broadcast. Darius came by, introduced himself, and offered us a golf cart to get around the fairgrounds. Sure! We hopped in his cart, went about 15 feet, and it died. Ran out of gas! Undaunted, he quickly hustled us onto another golf cart, which made our job so much easier. That was an example of what Darius is great at: helping people.

The mistake I made was telling Darius I took German at Purdue.

He speaks it fluently and when the girls were younger, he would leave a voice mail message on our home phone in German. A rabid University of Illinois supporter, he spoke to me in German as often as he could, and would wait for the answer to come back in my Purdue-educated German. And of course, I was never able to do so successfully.

Darius and his wife, Lois, have had several exchange students over the years. Young men would come from farms in other countries, and they learned a lot about farming. He stays in touch with them, and in the middle of the night here, he'll be on the phone to Germany and Belgium with them, asking them how they're doing. If they needed a piece of equipment to farm better, Darius would load up a container of stuff and ship it to them. And sometimes there were things he wanted, like antique IH equipment that the guys would find in Europe, and they'd ship it to him. He's very astute internationally and always knows what's going on in the world.

Darius may come across as a slow-talking good ol' boy, but he's smart as a fox and very insightful. Sometimes he's very subtle in his suggestions. For example, he'll tell you that you need to call someone by asking, "Have you talked to so and so?" If you hadn't and you knew you needed to call them, he was very persistent and very hard to turn down. And that's why the Half-Century and I & I shows have been so successful. People can't say no to him. When he asks, if he needs something, he'll "appreciate it if you'd do it."

Darius is the genuine article. He's kept me company on so many long rides in the car. He's a very perceptive guy. It's kind of eerie. When he knows I'll be behind the wheel for a long time, he'll call and we'll talk for a little while. And then he'll say, "Well, I'll get back to you in a little while." And when he calls back, he knows exactly where I am. I'll be driving down the road from Chicago to the home place in Indiana, and he'll call and say,

"Well, I guess you're coming into Stink City (Terre Haute) right now." And, by golly, I was just coming up on the city limits! He has a superb grasp of geography and about farming throughout the United States.

Darius was very involved in the restoration of Orion's tractor, and in fact, the restoration wouldn't have happened had it not been for him. Orion's sister, Norma, sent pictures of the tractor carcass sitting out in a pasture. And it was just a piece of junk. I arranged for Larry and Ben Eipers to go pick it up. Darius and I wondered if that thing could be restored. But the idea of the dean of farm broadcasters, the most respected and best-known farm broadcaster in America having the tractor he grew up with, well, that would be a tribute to Orion and IH.

At one of the early tractor rides, we arranged for that old thing to be there and we auctioned it off to benefit the Ag in the Classroom program. This program distributes teaching materials to rural communities

Orion, with grandson Matthew on his lap, and son, David, beside him, shakes hands with Darius at the unveiling of the rebuilt Samuelson family F-20.

and big city schools. Real life examples of how food and agriculture are used are part of the math, science and reading curriculum. It's a part of the Farm Bureau program and I've been very supportive of it.

The IH Collectors Club in Central Illinois, Chapter 10, bought that old carcass, paying over $7,000 for it. $7,000 for that hunk of rust! I think Darius still gets ribbed about it, especially by Lois. And then the restoration took place. Obviously, they had to get new parts, but did their best to keep whatever they could from Orion's dad's tractor. Had it not been for Darius bringing it all together, that tractor would not have been rescued and would not have been restored. Seeing it unveiled, with Orion, Gloria, David and Katherine there that night was amazing. There've been very few times I've seen Orion cry, but he did that night. Sitting there, with his family, on his dad's tractor.

Darius would acknowledge that he's a retired farmer now. His boys, Dirk and Derek, are both very good farmers and are both like their dad in that they are sharp operators and have learned from Darius how to farm well. He helps them at harvest and planting times, but he's limited to how much he can help them because of his age and his health. Along with his family, old International tractors, his passion for agriculture and his never-ending loyalty to the University of Illinois, Darius dearly loves his church and its German Lutheran congregation. Immanuel Lutheran Church sits at a rural intersection marked as Flatville. Beautiful, ornate and maintained in pristine condition, it's the church in which Darius grew up. He calls it "The Cathedral In The Cornfield." It has a prominent steeple and the flyboys at the old Chanute Air Force Base would use it as a marker when they were circling to line up and land on the runway.

One of the main ways the we bonded is through tractors. There's a lot of emotion tied up in these old machines. We spent a lot time on those tractors in our formative years with our dads, moms, grandparents and uncles. For a lot of people, a flood of memories are connected to these old machines: first dates, football games played, and days you're falling asleep on the tractor because you stayed out too late the night before. Of course, you were admonished by your father and mother for doing so. I remember working the Wabash bottoms by myself, hour after hour cultivating beans, never seeing another living, breathing human being all day long. Dad would get me started in the morning and he'd head back up the hills to work. It was a struggle to stay awake.

Then one afternoon, I was startled. People have often asked if there was a John Deere that was near and dear to my heart. My red tractor fans might find this hard to believe, but there was. I was about 14 years old and was pretty partial to a John Deere 4020. I was cultivating beans and coming up to an end row. Hot and dry, hadn't seen anyone for hours. And I look

The IH Collectors Club drove their Farmalls through the Rock Island IH plant before it was razed.

up, and there were two brand new John Deere 4020 tractors coming down the road. Herman Wilson's brand new 4020s driven by Herman's teenage daughters. In their swimsuits! Later on, my dad came down and asked what happened to those end rows that were torn out. "I don't know, Dad," I lied. "It looks like a bad case of insect infestation." He looked at me and responded, "Insect infestation?

I think you fell asleep." I didn't dare tell him the truth. All these years later, you may see me at an antique tractor show, eyes glazed over, looking off in the distance, staring at a pair of 4020s, recalling days of my misspent youth.

We went back to the old Farmall tractor plant and did a story, driving our old tractors that had been manufactured there down alongside what had been the assembly line. It was an amazing, historic day for us, thinking of all the tractors that had come out of that plant, including the ones we were driving at that moment. We went to the locker room at the plant. WGN's own Spike O'Dell is in the video. As a young man, he had worked as a security guard there.

I never named my tractors. Some have and you'll see the name printed on the side. I just called mine the "Super H," "Super M," or the 560. The 560 is stored at Darius's place in Champaign. That was my dad's tractor. It had been sold to a neighbor. Again, Darius played a big role in the restoration. He lined up the Prairie Central FFA chapter to restore it under the direction of the I & I Club, and helped find some of the parts. We had parts donated from five states. Nobody wanted credit for the parts they donated, they just thought we should have them on the tractor.

Darius and I have talked so much and have both been involved in the Half-Century of Progress. It was the farm equipment and the love for IH and the company's heritage that brought us together. And I've often said, it may be the equipment that brought people together, but it's the friendships that keep people coming back. I think that's the case with us. When you see folks who are on a tractor ride with you, or you're with them at a parade, it forms a bond and it's nice to come back each year to see the familiar faces. And it's always a wake-up call when you lose somebody and when you come to the show and they're not there.

It's a neat avocation, a fun hobby with its roots in agriculture.

Many of us are old farm boys, but that isn't a requirement. Some members had an uncle or grandfather who farmed and that stoked their interest in the tractors.

During our last lap through the Rock Island Farmall Plant in the winter of 2005, photographer Mike Hood lit the scene superbly, and the red paint on the "Super M" really came to life.

The wrecking ball arrived not many weeks after we were there. Several of the best minds in the I.H. Collectors Club Chapter 10 tried in vain to save the mural. There was just no practical way to do it.

Chad Colby

EVERY TWO YEARS IN LATE AUGUST, thousands of people converge on the old Chanute Air Force Base in Rantoul, Illinois, for the "Half Century of Progress," the biggest vintage farm equipment show in America. It is an amazing display of Old Iron, with daily demonstrations of plowing, planting and harvesting, all done "the good old way" with implements powered by hay, steam, gas and diesel.

There are tractor pulls and horse pulls, and just about anything you can imagine that can be done with these "living" reminders of our pasts.It has been a thrill for me to be a small part of these shows. There have now been seven shows and the attendance keeps growing, as tens of thousands attend from as far away as Australia and England. Chad Colby's drone photo, above, was taken at the 2013 show.

— *Chapter Twenty* —

Ol' Man River

One of the most enjoyable and memorable stories I have done for our farm television audience involved a five-day ride on a Mississippi River grain towboat. They are called "tow" boats, but in reality, they push. Towboats push barges lashed together to form a "tow." Cameraman Bob Varecha and I took a ride on the Sally Archer, operated by American River Transportation Company (a unit of ADM). And this was not a little tourist excursion with a barge or two. The boat itself is 200 feet long and was pushing 35 barges — five across and seven deep, three football fields long — down the Lower Mississippi, south of where the Ohio meets the Mississippi. Farther north, on the large rivers with locks such as the Ohio, Illinois and Tennessee, 15-barge tows are more common.

The Sally Archer's engines had a total of near 9,000 horsepower. It was deafening back in the engine room without hearing protection. But up on the front of the tow, all you heard was the water against the lead barges. As I sat there in a lawn chair with my life jacket on, just a speck in the towboat captain's field of vision, it was very, very peaceful.

Farmers have fought hard for river system funding because the navigable waterways are so crucial to the movement of grain, fertilizer,

Herman T. Pott National Inland Waterways Library

The Sally Archer on the Mississippi River at St. Louis

coal, oil and road salt. And think about this: To fill just one grain barge (52,000 bushels), it takes 13 jumbo covered railroad hopper cars or 58 semi-trailer trucks of grain. On our trip, the Sally Archer had the capacity to push the equivalent of 2,030 semi loads of grain! It is a very safe and efficient mode of transportation, and is far and away the most environment-friendly mode of moving our farm products.

Oh, I need to tell you about Mildred. She was the cook, and it was her last trip. Mildred was retiring after her journey with us, and she pulled out all the stops. "Boys," she said, "I have sandwiches, cake, pie, brownies and cold milk. The refrigerators are open all night!"

One other member of the Sally Archer crew I found fascinating was the captain. He had no hands. It was remarkable that he could steer that massive vessel and its giant rudders, precisely guiding the huge tow using just his forearms. The thing that struck me as we were going under the Vicksburg Bridge was the small margin of error. The navigable channel between the concrete piers of that bridge is only 846 feet wide. And here was this guy guiding a 250-foot wide tow, adjusting on the fly for the current, and doing it all without hands!

Another person who comes to mind along that same line was the late Don Vogel, an announcer who worked at WGN Radio in the late '80s. He ran his own control board, even though he was blind. The station put some Braille labels on the control board, and if you looked over his shoulder at the VU meters measuring his volume, the levels were PERFECT. Without sight! His hearing was so acute he didn't need to see his meters. I understand that the word handicapped is no longer in favor, but at the time I met these men, it struck me how handi-capable they were.

THE NATIONAL FARM MACHINERY SHOW in Louisville is a great place to greet old friends. Wayne Page, an aerial applicator from Rochelle, Illinois, always makes me smile. (Look at him; how can you not smile?)

About five years ago, I mentioned on WGN Radio that I had two instances in 18 months where I had hit deer: one near DeKalb, Illinois, and one near Owensville, Indiana. Well, Wayne called the station and said, "Hey Max! That was nothing. I hit a deer with my plane!"

Returning from a crop-dusting flight, Wayne was surprised by a deer that hopped up out of the weeds along the grass strip. The deer caught Wayne's spray boom as the plane landed. He missed a half-day of spraying farmers' fields getting the boom fixed ... thanks to Bambi.

MY DAUGHTER, KRISTI, WAS 10 when the photo was taken in 1996. Kristi and her daddy met Little Jimmy Dickens backstage at the Grand Old Opry. I remember being concerned about being in the way and not wanting to be a problem as the artists came and went. Kristi and I quickly realized that was the last thing we needed to worry about. They — especially Jimmy — made us feel so welcome.

Jimmy's kindness has never been forgotten by either of us, and we joined his millions of fans in mourning his January 2, 2015 passing at the age of 94.

Dea Hansen at the Chicago Board of Trade

— *Chapter Twenty-One* —

Moving the Markets

Back in the day when the trading pits of the Chicago Board of Trade Building were jammed and raucous, people from around the world stood in the glass-walled Visitors' Center overlooking the pits and heard Dea Hansen describe how the world's grain prices were being set at that moment down below.

Dea was a big fan of the radio station and was always so kind to Orion, Lottie and me. And she was just as welcoming to the farmers on the tour buses from places like Bean Blossom or Bone Gap.

After some 150 years of colorful open-outcry trading, the trading now is nearly all electronic. But 30 years ago, Dea had a lot to show the guests at her great Chicago tourist stop, and it really was one of the places to see in Chicago, right up there with the observation decks in the tall buildings.

Among the LaSalle Street milestones:

-1848: The Chicago Board of Trade was founded.
-1865: Grain futures are invented.
-1870: The first "pit," an octagonal trading space was created. The

name is misleading because rather than being lower than the rest of the floor, they are higher, with a few steps leading up to the "pits."

-1885: The CBOT constructs Chicago's tallest building at LaSalle and Jackson. That building was torn down and replaced in 1930 by the current building, which is topped by a three-story tall sculpture of Ceres, the goddess of grain.

-1898: The Chicago Butter and Egg Board opens, which became the Chicago Mercantile Exchange (CME) in 1919

-1969: CBOT expands beyond agriculture, trading in silver, and in 1972, added foreign currencies.

-1987: The beginning of the end for open-outcry: CME begins developing an electronic trading platform called Globex. The first electronic futures trades came in 1992.

-2002: CME and CBOT merge into CME Group.

The 30-foot statue of Ceres, goddess of grain, overlooks the LaSalle Street canyon from atop the CME Group building.

This summer's closure of many of the open outcry pits sparked a flood of memories of trips to the floor of the Chicago Board of Trade. Names come to mind of guys who were hoarse when you spoke with them, voices raspy or gravelly from all of those years of yelling at each other as they stood in the pits elbow-to-elbow and face-to-face, flashing hand signals that resulted in thousands, if not millions, of dollars changing hands. There are many colorful, hilarious, and sad stories about the traders' antics, and the toll that the hard lifestyle took on some of them.

Don Pflaum, Kenny Gorgal, Dave Keefe, Howard Stotler. These were some of the guys who would step out of the pit to shake our hands when Orion and I would come down to the floor. Many of them listened to us on their way to the train in the morning or as they drove to the exchange at LaSalle and Jackson.

One of the first traders I met was Ken Gorgal. He was a corn trader who had no trouble holding his position on the steps of the pit, even on the most raucous days, because he had played football, both in college at Purdue, and professionally with Green Bay and Chicago. I had lived across the hall from Kenny's son at Purdue.

There were times when Orion and I moved the market. It happened on a couple of occasions and it just reminded you how careful you had to be about what you said. For many years, Wally Phillips's WGN morning show ratings were bigger than WBBM and WLS combined. You'd see a comment from a wire service provider and you'd realize it was something Orion and I talked about. On a hot summer morning in 1983, as traders worried about crop damage from drought, I went down to the floor right after the open and asked a veteran trader about the rally that morning. I knew who he was, but he didn't know me. I asked, "What's with this rally? What's behind it?" He said, "Orion Samuelson was talking on the radio this morning about the chickens dying in Arkansas from the heat." Well, Orion at the time was thousands of miles away from the microphone, traveling in Asia. I had been the guy on the air that morning with the story about croaking chickens. It was MY voice on the radio that helped move the world's grain prices that day!

But for many years, NOBODY moved the markets like Tom Skilling. That was demonstrated on a grand scale back in the pre-Internet days at the Chicago Board of Trade. There was a time near the end of the pit trading when the traders watched Tom Skilling's forecast everyday on WGN-TV. They put a big screen TV on the floor to show his forecast at 12:45 p.m., and there were many days that the market shifted between 12:45 and 12:50, responding to Tom's weather report. He had an immense sway on the market. There was even a rumor that Tom's wife traded the markets and Tom wasn't married!

It was amazing to watch Tom move the market, but traders could be swayed by just looking out the door. There was an expression that if it rained on LaSalle Street, the traders thought it rained on the whole Corn Belt. Orion and I both saw this happen. During the drought of 1988, when it was very dry in the Corn Belt, rain over downtown Chicago dropped the grain market immediately. Corn fell a nickel a bushel, and beans dropped 15 cents. And this was all on the basis of the rain falling downtown. Orion and I just looked at each other and shook our heads.

— Chapter Twenty-Two —

Hoarse Traders

C ommodity traders often wound up in the office of Dr. Robert Bastian. He was a voice doctor who took care of all sorts of people who wore out their voices; singers, professors, and even a couple of broadcasters I know.

In the early '90s, I was hoarse. I had gone through a whole year of being unable to clear my throat or find my voice. Generally, I recovered in 24 hours or maybe a few days. But this one had gone on for weeks and I had a raspy sound.

I went to an Ear Nose and Throat doctor who took a look and said, "I need to get you to an otolaryngologist. Go see Robert Bastian over at Loyola." I did, and he stuck a camera down my throat and had me sing "Happy Birthday." Not only did he discover I wasn't much of a singer, he found a nodule on my vocal cords. "Well, Max," he said, "I've got good news and bad news. The good news is I can fix this. The bad news is you'll have to be off the air for three weeks." That was in September and the difficulty was finding a hole in Orion's schedule where he wasn't traveling and could cover for me. It was the following June before we were able to find a gap in the schedule.

So throughout that winter I was rather raspy on the air, which was a blessing in disguise. A commercial producer called me at the station and asked me to do the voice-over on a TV spot they were making for Ford pickups. I apologized, telling him my voice wasn't the greatest at that moment and was raspy. He said, "I know! I hear you every day and that's exactly what we want for this spot." I did the commercial and it ran the night of the Winter Olympics. It aired between Tonya Harding and Nancy Kerrigan skating. There was so much attention on the fiasco involving those two. It was a huge night for viewing. One skating performance was scheduled before eight p.m. and the other was after eight. My spot ran about a minute till eight. One of those fun things I never expected to do, but my raspy voice gave me an opportunity!

After Dr. Bastian removed my nodule, it was such a feeling of relief. I felt as though I'd been made whole. He gave me my career back, and he eventually helped Paul Harvey, too. About five to eight years before he died, Paul was hoarse and couldn't get any help. He even went up to Mayo and they weren't able to help him. He ended up going to Bob Bastian, who at that time was at Good Samaritan. I remember being on the air at WGN the day Paul came back, and he mentioned the good work Dr. Robert Bastian had done. You really could hear the relief and appreciation in Paul's voice, having been made whole. I knew that feeling.

Chapter Twenty-Three

Joan Rivers

W hen Joan Rivers was still alive, I told people about meeting her. I joked that it was "back about 15 facelifts ago for Joan." It always got a good laugh, but I was never serious. I mean, who could have had that much plastic surgery? Well now we learn from Joan's daughter, Melissa, in her book about her late mom, that Joan was never happy with the way she looked, and actually had 348 plastic surgeries!

Our meeting took place at the tony Chicago restaurant Spiaggia on the northern tip of Michigan Avenue. The restaurant overlooks Oak Street Beach on Lake Michigan. We were there for an event hosted by Tribune Entertainment, the television production arm of Tribune Broadcasting. Networks frequently throw lavish parties where they trot out their talent to schmooze with advertisers and media buyers, but this was one odd mixture of "talent."

The "personalities" of Tribune Entertainment who were in attendance included Geraldo Rivera. Yep. Remember "The Mystery of Al Capone's Vaults"? It was produced by Tribune. Also there was the late Don Cornelius of *Soul Train* fame. Dennis Miller was there, who hosted the

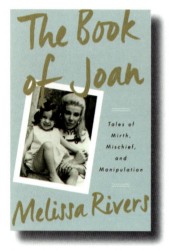

short-lived *The Dennis Miller Show*.

And then there was Joan Rivers, whose late-night *The Joan Rivers Show* was produced by Tribune. Joan was quite pleasant and gracious, and, despite how she apparently felt about her looks, I thought she was pretty.

Also in attendance were two farm broadcasters, Samuelson and Armstrong, who did Tribune Entertainment's farm show.

Yeah, what was wrong with THAT picture? It was funny, and fun.

After Siskel and Ebert left Tribune Entertainment and took their movie review show to Disney, the show continued with different hosts. I continued as the announcer, and what a difference it was to say, "At the Movies with Rex Reed and Dixie Whatley."

"AND NOW, FROM THE WIRES of United Press International ..."
(Sure, a <u>radio</u> guy needed a tie, a white shirt and a three-piece suit
— to look serious for a photo shoot.) This was in 1975 at the Illinois
Farm Bureau, where I was Broadcast Editor for two years, just before
joining Orion in Chicago at WGN. The machine, an Extel, was prim-
itive by today's standards, but it was a far cry better than the noisy old
belly-high teletypes that shook violently as they typed.

YOU HATE TO PART WITH THEM. They've been a part of your life for a long time. Pretty? Once, maybe, in the way that brand new boots are, but not anymore. People may make fun of them and ridicule you, but you don't care who sees them and you don't care how they look. They're you. They hold memories. You use and abuse them, you kick things and try to force things into place with them.

My most recent pair of well-worn boots are shown above. The soles are gaping away from the shoe, but they've still got a lot of miles in them, and they are soooo comfortable. I've had old boots that fell apart while I was wearing them. You can spend a lot of money on boots, but I've had the best luck with the less expensive ones. Oh, I've had expensive boots, thought they were nice boots, but I found them tough to break in and they didn't always feel right. Some of those boots spent a lot of time in the garage and not on my feet.

At the 1975 Illinois State Fair. Governor Dan Walker might be saying, "Hey kid! Lie down on the ground so we can play chess on your shirt!"

I'm not sure why this farm broadcaster needed a police ID, but I was given one. Maybe just in case the American Agriculture Movement tried to take over the Chicago Board of Trade, or a cattle truck got caught going down Lake

CHICAGO POLICE DEPARTMENT
1988 NEWS MEDIA IDENTIFICATION
Max Armstrong
is a duly accredited
REPORTER   of

WGN RADIO

LeRoy Martin
Superintendent of Police
Expires
March 1,   No. 01733 N
1989

Shore Drive, or an intruder with a step ladder showed up (as happened more than once) at the Farm In The Zoo in Lincoln Park.

This is a Tasmanian Devil farmer I met in Australia. Because of disease, the Tasmanian Devil has been in danger of becoming extinct, and Colin Wing, also a beef producer, takes in injured devils, nurses them back to health, and shows them to tourists who visit his farm/preserve. He has captured a real agritourism opportunity. (No, that is NOT the devil on his right! That is one below us, however.)

I'm shown in WGN's Studio B at Tribune Tower, where Eddie Schwartz worked at night. "Eddie Schwartz, news and sports. Call in to say hello. You're in tune with Chicago night-time radio."

The studio was new when this photo was taken, and Eddie hadn't yet scratched the console with the ruler he used to move the faders and push the on/off buttons for each channel.

— *Chapter Twenty-Four* —

Small World

Hardly a week goes by where I'm not amazed by the "small world" experiences that show up in my life. What follows are are just a couple of examples.

Recently, a lifetime friend (grade school, high school, Purdue) from Indiana was sitting with her husband in the Newark, New Jersey, airport, en route to a vacation in Scandinavia. They struck up a conversation with a stranger sitting next to them. He was flying home to England after several days in the States. Where had he been? In Decatur, Illinois, where, two weeks to the hour earlier, he had been on stage at the Farm Progress Show, being interviewed by ... me.

One hour later I got a call from another of my very best friends, an Illinois farmer. He told me that his neighbor, a farmer adopted at birth 40-plus years ago, had been trying to find his birth family and was able to track down his grandfather, but he had died a few years earlier. The grandfather was a farmer, too, only 100 miles away, and in his sleuthing, the younger man read a sympathy note in the funeral home on-line guest book. It was written by ... me.

In a phone conversation, I was able to share with the grandson how bright, personable, funny and caring his grandfather was, and how lovely and sweet his grandmother is. I also shared with the younger man a photo from my files of his grandparents.

This farmer-grandson says he has been so blessed with a great family and a wonderful farming career in a tremendous community. He has never wanted to travel anywhere or do anything else. He loves his life! His voice cracked as he told me the only thing that had been missing was the connection with his blood relatives.

We reach out with care, knowing how delicate these situations can be, but I continue to be amazed — and humbled — that my small world can provide some sort of connection between others in a very BIG world.

Chapter Twenty-Five

A Dangerous Way to Make a Living

W e've spent a great deal of time over the years spreading the good word about agriculture and sharing the admiration we have for the great work our producers do, but we've also spent far too much time spreading the sad news about farmers and ranchers who have died or been seriously injured in accidents. It's no surprise, when you consider that our jobs involve heavy machinery and unpredictable animal behavior. The federal government ranks agriculture as a more dangerous occupation than mining, and the risk of death in farm jobs is five times higher than all job categories combined.

Tractor accidents are atop the list, and in recent years, ATV accidents have become more common. PTO accidents, where a piece of clothing or something gets caught in the power take-off shaft that's spinning at over 500 revolutions per minute, maim and kill farmers.

Any farm implement is a potential killer. A few years back when I met celebrity and farmer Charlie Daniels, he told me about his nearly catastrophic encounter with a post hole digger. Charlie was digging fencepost

At the DuPage County Fair in 2002, Charlie Daniels told me the farm accident that nearly ruined his career "happened in the blink of an eye."

holes on his farm near Mount Juliet, Tennessee. His glove, then his shirt sleeve, got caught in the auger. The spinning blades pulled him in, broke his right arm in three places, and broke two fingers, too. That wasn't good for a guy who makes his living playing guitar and fiddle. Charlie said he told the surgeons, "Please restore me so I can play again." The injuries required surgery and sidelined him for four months, and Charlie felt lucky to get off that easy.

Too often, the stories I've reported involve children and teenagers. There was a terrible accident recently in Eaton County, Michigan, where a teenage girl was putting bales of straw on an elevator up into a loft, got caught in the PTO, and died instantly.

When it comes to children, we can all agree that one injury or death is too many. But there's a lot of disagreement about how to regulate child safety. The U.S. Department of Labor floated some ideas a few years

ago, but farmers and farm organizations fought the interference and the pro-
posals were dropped. There's ongoing concern that OSHA or one of the regu-
latory agencies will step in and erect all kinds of barriers that would prohibit
kids from playing an active role in the farm. And the problem is the regulato-
ry agencies have a tendency to go way too far. The fear is that the regulations
would be so rigid that kids would be prohibited from working with livestock
for 4-H and FFA projects. It's a complicated topic because we need the labor
of young people on our farms, and young people on farms need the labor to
help them mature. Many of us look back and think, "Jeepers, how much we
would have missed growing up if we had not been able to work alongside our
fathers, grandfathers, uncles and mothers!"

I remember a phone call into WGN Radio from a farm mom calling from a combine cab, and you could hear a baby in the background. I
asked, "How old is that baby?" She responded, "Oh, he's about six weeks
old." The truly rigid safety experts say to not take the child to the field and
don't let them around any equipment, and don't allow them on the tractor.

I'm sensitive to that. I don't want to encourage kids to hop on a
tractor at a young age. It's a difficult discussion because the kids want to
be there with their parents, and the parents — who were on tractors at
a young age themselves — want them to be there. There are extra seats
in some of the cabs that allow the parent to bring along the child. It's a
wonderful bonding experience that a young person never forgets. All of us
remember those experiences.

We like to think that farm kids grow up learning about and under-
standing the dangers, and are kept away from risky chores until they're able
to do them safely. That is the case, for the most part. In fact, the government says agriculture-related injuries to workers younger than 20 dropped
by nearly half from 2001 to 2009. Farm injuries were most common among

children ages 10 to 15, but they also dropped by nearly half during that period. We're improving, but we need to do better.

Being in this business as long as I have, I've met a number of families who had a tragedy of some sort. And when it does happen, it hits the whole community. Seeing the crushing grief, guilt and heartache these accidents cause is one thing you never want to experience again.

Grain bin accidents take many lives each year, and as more storage bins are built on farms to handle the higher yields, there are more accidents where people are trapped under tons of grain. Ohio State University Ag Safety Specialist Dee Jepsen says it can take only seconds for someone to get stuck. And death can come quickly. "It's just like a boa constrictor around you," she says. "That grain is very heavy and the further down you

We could use more billboards like these.

I had the opportunity to meet Dr. Dee Jepsen and tour OSU's mobile grain bin safety exhibit at the Ohio Farm Science Review near Columbus.

go, your lungs aren't able to breathe and it truly is a suffocation." She says the vast majority of people who are trapped do not survive.

I had the chance to meet Dr. Jepsen and see the mobile grain bin that Ohio State ag students built as part of an effort to reduce fatal accidents. It's on a 40-foot trailer and it travels around Ohio conducting training exercises with the state fire marshall for local fire departments and first responders. They demonstrated to farmers how a person can be pulled down so easily in the grain. But I've seen these demonstrations for over 15 years and the bottom line is people are still getting trapped in grain bins.

It isn't just grain bins that can swallow you up. Silos, manure pits and bulk tanks can be just as deadly.

And do not ... I repeat ... DO NOT take a rural intersection for granted. Unspeakable tragedies happen every year in late summer when the tall corn creates "blind" intersections where the view is obscured. My farmer friends have their own accounts of close calls or of seeing people just blow through intersections in front of them. The fact is that in farm country, many rural gravel and dirt roads do not have stop or yield signs.

Some that were in place at one time may have been knocked down.

And the thing about nearly every farming accident that I can't get over is that they were preventable. Be careful. We need you.

One of those dangerous, blind country intersections

— Chapter Twenty-Six —

Tim Alexander
Winamac, Indiana
Pulaski County

*I*n 1993, on a beautiful September afternoon, Tim was working on the inside of a grain bin, his ladder leaning on a stirator. He fell about 15 feet and his neck landed on the ladder. Tim was paralyzed from the shoulders down with a C3/C4 spinal cord injury.

It was right at harvest time. I got a call from one of his neighbors, Karen Fritz, and she said Tim and Sue's neighbors would be pitching in to harvest the Alexander crop. She just wanted to let me know. I talked about it on WGN that morning, and got a call a short time later from Hugh Hill, a longtime Channel 7 political reporter, who'd heard my report. Hugh typically covered mayors and governors and spent a lot of time in Springfield. He asked, "Where is this? I'm tired of chasing the politicians. I'd like to get out of town on this beautiful day." I gave him details and directions, and I recall he took a helicopter down and shot aerial footage of these guys harvesting.

About 10 years later, I ran into Hugh at a White Hen Pantry in Naperville, and he said, "One of the most memorable stories of my 40-year

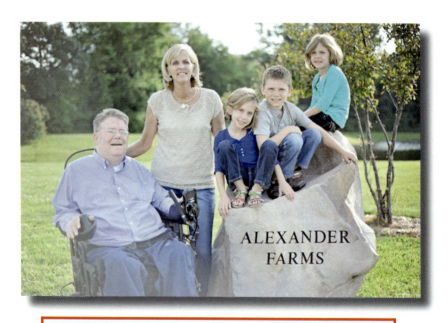

Tim and Sue Alexander, with three of their grandchildren

Legendary Chicago TV
newsman Hugh Hill

career is the one I got from you, the one about helping that farmer in trouble. I had never seen anything like that, the way farmers put everything on hold to help each other."

It actually happens more than we know, and has gone on for decades ... every spring planting, and every fall harvest. Usually, you don't hear about it until after the fact. Guys just show up and do it. They don't call attention to it, nor do they expect any. "There, but for the grace of God, go I," is how many have explained it to me. "I could be in a tough spot, too. And if I had gotten into trouble, this guy we're helping would've been the first one in line to help me out."

It is a stirring sight when farmers put everything on hold to help a neighbor, and all at once, with 15 to 20 combines, 30 trucks — whatever it takes — they get the job done in a hurry. About five years ago, I was out on a ridge in Dekalb County and saw 18 combines moving at one time, a cloud

When Iowa farmer Gene Sitzman died in a combine accident in 2014, neighbors gathered the next day to finish his bean harvest.

of dust accompanying each one. And when you see all those trucks lined up along the road, and you know they're there for a neighbor who needs a helping hand, you never forget the sight.

One snowy night, I went to see Tim at the Chicago Rehabilitation Institute in downtown Chicago. He was alone in his room and I recall there was a basketball on the shelf that had been signed by the Purdue basketball coach, as well as a lot of other Purdue things around the room. As is always the case when I see him, Tim was smiling. He would have a lot of reasons to not be smiling, and anyone could understand someone withdrawing and being mad at the world. But Tim Alexander has never been that way. All of his neighbors have said what a remarkable guy he is.

At Brookston, Indiana, 30 semi-trucks lined up as farmers gathered to harvest crops for a deceased neighbor's family. Walter Chitty always helped others, local growers said.

Lafayette Journal Courier

Just as remarkable is Sue, his wife, who shouldered a huge amount of the responsibility of keeping the family and farm going and growing. Sue was honored at the Indiana State Fair in 2006 with the Women in Agriculture Achievement Award.

At the time of the accident, the Alexanders had over 1800 acres. Amazingly, and against huge odds, they've continued to grow. Tim operates the keyboard and his wheelchair with a straw-like tube. He's got a van he can get in and out of. He and Sue are just tremendous farmers and they are a great team, an asset to the farming community. They are two of the most inspiring people I know.

I've asked Tim to say a few words:

Thank you, Max. Yes, we have been very fortunate and have slowly grown to our present 4,000-acre size. That seemed very unlikely to ever happen when I was in Northwestern Memorial Hospital's intensive care after the accident. I was unconscious in a controlled coma for a month. Some pretty amazing things happened while I was under. A couple of our neighbors spread the word about what they were planning to do, more people started volunteering and, as Max mentioned, dozens of our neighbors, friends, and even people we didn't know well, got together and harvested our beans in one day!

By mid-November when the corn was ready, our neighbors were ready to do it again. I was conscious and wanted so badly to be home on the day of the harvest. The doctors didn't want to let me go, but they eventually agreed that I could, if I brought a nurse with me. One of our friends picked us up at the hospital and I was at my farm by noon.

More than 100 friends and neighbors came with 50 combines, 20 grain carts and 100 trucks. By two o'clock they were finished harvesting our 1,000 acres of corn. It was so emotional for me. Everybody joined in, and the

*women cooked for everybody. The elevator was shut down that Saturday —
except to receive our corn. I made it over to the elevator and saw a long line
of trucks with only our corn going in. It was such a heartwarming moment
and we were so shocked, yet thankful, for all the work being done. The impact
it had on us was immeasurable.*

*Sue and I wouldn't have been able to keep the farm going and grow-
ing without a lot of help. Sue's dad has been a tremendous help. He was my
mentor as Sue and I began farming and still helps us even after he retired.
He's a great hands-on guy, and, of course, I couldn't do the hands-on work
anymore. Also, Wade Shedrow, who had just started working for us and
was at the bin with me when I was hurt, came back and worked for us full
time. He was only 21 at the time of my accident and has stayed with us all
these years. He's become part of our family. Twelve years ago, we hired Jason
Alma, a second full-time guy. They are all you could ask for as employees
and, together, they have helped make it all possible.*

*Our landlords were very supportive, too. When I was in the hospi-
tal, Sue and I made the decision that we wanted to continue farming, if we
could. Sue visited each of the landlords and asked them if they would support us. Every single one of them did. As the years went by, opportunities
to expand opened up as other farmers retired and we were able to acquire
additional ground.*

*One evening while I was in rehab, around supper time, a nurse
came in and said I had a visitor. And lo and behold, here comes Max Armstrong! We had never met, but, of course, I knew who he was from radio, TV
and farm shows. I was just flabbergasted that Max Armstrong would come
to see me. He's been to visit our farm many times since, and we've become
good friends.*

When the accident happened, I was 36 years old. The kids were only 12, nine and four. They all went to college, and although they stand up to support farming whenever they get an opportunity, they chose other occupations. Megan is Brand Manager and Communications Specialist for BraunAbility, a leading manufacturer of handicap-accessible vehicles. Our son Jim is a professor at Franklin College teaching Religion and Philanthropy and Mark is an audio production engineer for Tour Design, a division of Live Nation Entertainment. They all come back quite a bit, and Sue and I have hopes for the next generation: we have a grandson who is very interested in farming!

(Author's note: The story of how neighbors rallied to help Tim and Sue became widely known, thanks to Orion Samuelson, Paul Harvey and Hugh Hill. Below is the transcript of the broadcast.)

Orion Samuelson: *On Samuelson Sez this week, we share with you a bittersweet story, a story that's all too familiar this time of year, where a farm accident causes injury or death that makes it impossible for the farmer to harvest. The upshot of the story is the neighbors come and do the job for him. That story unfolded this week around Indiana farmer Tim Alexander. Our Dave Russell and Max Armstrong started it on our WGN Radio show. From there it was picked up and carried on radio worldwide by Paul Harvey and on television by our friend Hugh Hill on WLS-TV in Chicago.*

Paul Harvey: *Agribusiness? You heard Max Armstrong report one week ago, Indiana farmer Tim Alexander was working in his grain bin near Winimac. He fell from a ladder. He is undergoing surgery right now in Chicago for a spinal injury. But before he went under the anesthetic this morning, they got word to him that 80 of his neighbors had converged on his farm with 40 combines and 30 trucks, and his 800 acres of soybeans will be tended to today. And if Tim is not up and about by harvest time, they'll take care of his corn, also.*

Hugh Hill: *The combines chewing through that Indiana soybean field say simply this: "When your neighbor is in need, help him." Tim Alexander, 36, with a wife and three children, lies tonight at Northwestern Memorial Hospital in Chicago paralyzed from a terrible fall from a ladder. Back at his beloved farm near Winamac, his neighbors are harvesting his beans. Mike Blinn is one of them: "It's one of those things that has always been, the same way grass is green, when your neighbor is in trouble, everybody gets together and helps out."*

After Orville Redenbacher was there (and had gone on to put his namesake popcorn in a jar), Princeton Farms in Princeton, Indiana, was canning its popcorn. The can above dates back to about 1972.

With this husking peg, my dad hand-harvested 100
bushels of corn in nine hours. Modern combines
do it in nine minutes.

— *Chapter Twenty-Seven* —

The Husking Peg

W hen you look at the photo on the previous page, it's a little difficult, if you've never seen a husking peg, to imagine how it was used. In a simple description, the fingers of your right hand (for right-handers) fit into the holes on the left, with your thumb free on the right side to guide the peg along the length of the corn cob, separating the husk from the cob. The method most common was to grip an ear with the left hand, yank off the husk with the peg in the right hand, and then, with a quick twist of the cob to remove it from the stalk, the husker tossed the liberated cob into a wagon parked a few feet away. On the wagon, the opposite side from the husker had a higher sideboard to lessen the chance that the husker would overshoot the wagon. That sideboard was called the bang board. I assume that name came from the sound of the corn cob smacking against it before falling into the wagon bed.

The average farmer could husk about 300 ears in 80 minutes, which was the length of the contest. Of course, there were some who were faster, much faster, and husking became a competition across the Midwest. Much credit for this goes to Henry Wallace, who, in the 1920s, was editor of the *Wallaces Farmer* newspaper, which today is part of the Penton Farm

Progress media family, my employer. He went on to start the seed company which became Pioneer.

Nebraska Farmer, part of the Penton Farm Progress family, sponsored many of the events. A *Nebraska Farmer* magazine article I found on line told the story of how popular corn husking became. In the 1920s through World War II, husking was a great spectator event in middle America, outdrawn only by the Indianapolis 500. Tens of thousands of people would turn out to watch the fastest shuckers in the country. The 1936 National Cornhusking Championship attracted 160,000 fans. It was so popular that NBC decided to carry it live on its radio network. A portable transmitter was used so that real sound effects could be brought right to the listeners.

The contest was stopped because of the war in 1941, with plans to resume after the war, but it never got restarted, probably because of the

widespread use of mechanical pickers. Still, to see champion-caliber shuckers throwing 40 to 50 ears per minute must have been a sight to see!

A contestant pitches an ear of corn against the bang board.
Courtesy: Nebraska State Historical Society

Dave Wulff supplies sweet corn to some 50 restaurants in Chicago. From his farm an hour south of the city he ships three truckloads a week.

Yes, Wulffie is a John Deere man. He has the shirt, wears the hat, and on a hot summer night I recall seeing a John Deere tat. Nobody's perfect.

— Chapter Twenty-Eight —

Edmund Muskie

The only person to ever swear during my interview ...

My recollection of Edmund Muskie is a short one, but it stands out because he was the only person, of all the thousands of people I've interviewed over the years, who swore during the interview.

Senator Muskie, from Maine, was a presidential candidate in the '60s, was on the Hubert Humphrey ticket in 1968, and was briefly Secretary of State, serving the final seven months of the Carter Administration before President Carter lost to Ronald Reagan. As happened with many politicians of that era, Muskie's view of U.S. policy in Vietnam evolved, and years after the conflict ended, he tended to look at the Vietnamese differently.

In the late '80s, I had the opportunity to travel to Vietnam as part of a trade mission to look at it as a possible market for American farmers. After our return, we had the opportunity to have dinner with Ed Muskie. There must have been seven or eight of us around a table. When I approached him about doing an interview, he said, "Oh, you bet." We stepped outside and our cameraman, Phil Reid, fired up and started rolling. I asked Muskie about Vietnam and his view on the potential for economic development there. I recall the population at the time was about 72 million, and it was a young and rapidly developing country with probably 75 percent of the people under the age of 30.

During our trip, when we would bring up the subject of the war with someone we sensed might have fought with us or against us, it was the last thing they wanted to talk about and it appeared to be of little to no interest to them. They had long ago put the Vietnam War in the rear view mirror. The future and moving ahead was what it was all about.

That was Ed Muskie's focus, too. He had observed, as had we, how industrious the Vietnamese were. We saw people sweeping airport runways. With hand brooms! As I interviewed him, the more he talked about the Vietnamese people, the more excited he got. He gestured wildly with his arms and said, "They're so god damned industrious! You've got to love those people! They work their asses off!" He was so enthusiastic about the Vietnamese people, and they've become enthusiastic trade partners. In 2014, Vietnam imported more than $2 billion worth of U.S. agricultural products.

— Chapter Twenty-Nine —

Edward Telling
Father of the Discover Card
But that was the last thing he wanted to talk about ...

I first met Ed Telling in 2001 at our front door in Naperville, Illinois. He was new to the neighborhood, but did not need to introduce himself. I sure knew who he was. In my business reports on WGN Radio in the late '70s and early '80s, I had spoken his name many times when he was Chairman of the Board and CEO of Sears. Those were the years when Sears truly was, as its ads stated, "Where America Shops." In 1981, 70 percent of Americans had a Sears credit card.

When you think back to that era, Sears was larger than life. People in every community in America were touched by Sears. I remember when the Sears & Roebuck catalogue would arrive in the mailbox each fall. That thing was dog-eared and had pencil marks all over it by the third week of December. For earlier generation farm families, the two books you were sure to find in every house were the Holy Bible and the Sears catalogue.

As a child, I remember going with my dad to the Sears store in Evansville, Indiana, which is one of the stores where Ed worked on his way up the ladder. You'd walk in and see a guy demonstrating vacuum cleaners, and I can still remember being there with my dad, and the smell of the peanuts and the candy counter.

Edward Telling spent his entire career with Sears, 38 years from trainee all the way to the top — truly, the very top. From his office a thousand feet over Chicago in the Sears Tower, Ed ran the largest retailer in the world. When he took over, it was a tumultuous time for Sears with discount chains beginning to loosen Sears' grip on the market. Some retail industry analysts of the day were not very kind to Ed's leadership, but he turned Sears around. Ed is credited with helping Sears hold on as he led the company into the business of financial services.

In a remarkable 40-day span in 1981, Sears started a money market fund and acquired both Coldwell Banker, a real estate agency, and Dean Witter Reynolds, a Wall Street brokerage house. The Discover Card was created soon after.

Ed wound up at our door as a result of one of life's curveballs. His wife, Nancy, his high school sweetheart, died in 1996. His administrative assistant for many years, Lynn, lived at the end of our block. She had been married when working for Ed, but she and her husband had gone their separate ways. A few years later, Ed's wife died and he eventually married Lynn.

When I answered the doorbell, there stood a white-haired gentleman holding a paper placemat that he brought back from a restaurant somewhere near Danville in Central Illinois. Danville was his birthplace and hometown. He wanted to take Lynn's children to show them his roots. Ed had taken them into a small restaurant and saw the placemat with my picture on it promoting a tractor show. At our door, he held out the placemat and said, "I thought you might like to see this." There was another time when I was sitting on my deck and Ed was walking down the sidewalk with his dog, and the first thing out of his mouth was, "How's that tractor of yours?" We sat and really had a good visit and I asked him questions about Sears because I thought it would be nearest and dearest to his heart. This was less than 10 years after his retirement and his extraordinary corporate experience was not that far in his rear view mirror.

You helped give birth to the Discover Card, right? "Oh, yes, we had some very busy years, but you know some of my most memorable experiences were in the stores, as a younger man. I worked in Danville, and you know I was down at Evansville for a while. I worked with some fine people there."

Your strategy of diversifying Sears seemed to really pay off, right? "Well, we felt that we needed to sell more than just goods. You know, we sold tractors at one time, Max. The Sears Economy tractor was before me, and we had the little David Bradley walk-behinds, they called them. But frankly, we just weren't very good at selling tractors."

What was it like having an office at the top of The Sears Tower, with that amazing view every day? Well, Max, it's a long way from the road crews in Central Illinois. As a college student, I worked on the highway crews in the summer. It was a good learning experience, especially when I refused to join the union just for that

Danville, Illinois, has long been an industrial and agricultural center for Central Illinois, and has produced a remarkable group of favorite sons. Included are nationally known entertainers such as Donald O'Connor, Dick Van Dyke and Gene Hackman. Among the corporate world leaders is Edward Telling.

summer job. Someone took a shot at me while we were out there and I was hit in the shoulder. My dad took me into the doctor's office near Danville,

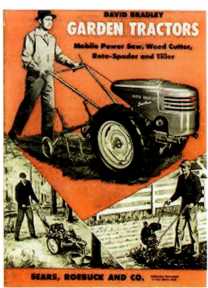

Sears was a big player in agriculture in the 1930s. It produced one- and two-plow tractors at plants in Peru, Illinois and LaSalle, Illinois. For many years, Sears sold David Bradley walk-behind yard tractors.

and they patched me up. A minor wound, it was."

I am not sure what it had been like to work for Edward Telling, but in the final years of his life, not long before his passing in 2005, I very much enjoyed our visits as he walked the neighborhood. I was humbled by the fact that he carried that placemat all the way back from Central Illinois. That he thought enough of me that he wanted to show this to me, I found remarkable. He was a quiet, friendly man, warm and cordial with nothing artificial or fake about him, and not a hint that this was a guy who

Max Armstrong

had run the largest retailer in the world. Ed Telling was one of those "Greatest Generation" men. He served as a Navy pilot in World War II before going to work as a Sears trainee in 1946.

> *It's so rare these days for people to stay at the same company throughout their lives. My grandfather, who was a radio telegrapher, was with the Chicago and Eastern Illinois Railroad for 56 years. He started with them and stayed until the day he died.*

What impressed me about Ed Telling? His humble roots and the way he worked his way up, from stock boy to running the corporation. More importantly, it was just his heartfelt interest in that smaller community, that rural area in Central Illinois. It reminded me that along this path we're on, we have all the zigzags of our lives with diversions and mishaps. Sometime down the line, many of us harken back to how we got it all started. Fond memories of the people in our early lives. And even not so fond memories. We realize how those people and events shaped our lives and allowed us to do what we did. So here was this guy who had been among the most powerful leaders of a giant corporation as well as a board member on other corporations. Here, at the end of the day, what mattered most to him was where he had come from.

When I asked questions about Sears, he didn't dodge them. He would answer, but they were short answers and it was quickly evident that his time at the helm of Sears was about the last thing he wanted to chat about. He wanted to talk about his roots. That is what mattered to him.

I'll never forget the day when this old farm boy was wandering, alone, through the official residence of Prime Minister Tony Blair. I suspect security has been tightened up since then.

Chapter Thirty

10 Downing Street

My unusual meeting with Prime Minister Tony Blair

"*I* have good news and bad news," said the woman from the British tourism ministry as she met us at the London airport. Okay, what is this about? We had just flown 4,000 miles to interview Nick Brown, the UK's Minister of Agriculture. She gave us the bad news first. "You're not going to meet with the Agriculture Minister to do the interview."

Well, we had come all the way from Chicago where, every night on the TV news, we watched with shock, sadness, and fear the reports of the UK's battle with Hoof-and-Mouth Disease, which included the mass burning of carcasses to stop the spread of the disease. At one point, 90,000 animals a week were being euthanized, far more than farmers and British agriculture officials could manage, so the British military had to be called in to help. Obviously, it was something we did not want in the U.S. We decided to go to London and do a story about the disease, explore how the Brits were handling it and what could be done to keep it out of our country.

Not only was it devastating the British livestock industry, it was killing tourism. Nobody wanted to go to the United Kingdom because they thought they would see piles of animal carcasses. So the UK was very interested in telling the story, but it was much more than just this agriculture story.

A woman at the British Consulate in Chicago helped arrange our trip. Cameraman Phil Reid and I expected that we would go from the airport to Parliament and meet with Agriculture Minister Brown. When our British host delivered the bad news that we would not be meeting with Mr. Brown, my mind raced to figure out a Plan B. Then she added, "Instead, you'll be going to 10 Downing Street to meet with the Prime Minister, Tony Blair." Well, okay. That was better than any Plan B I had come up with.

Phil and I arrived at 10 Downing Street, but only I was cleared to go into the building. The bobby guarding the entrance made it clear that Phil and his camera had to stay outside behind the ropes. As I went inside, Phil set up his tripod so that the camera was aimed straight at the door, but at that time the bobbies would not allow photographers to shoot in when the door opened. Phil, always looking for the best shot, was given a stern warning by the bobby. "No, you can't do that. Move on down." Phil reluctantly moved, then moved back again a few minutes later. That's when the bobby whacked him across the butt with his baton and told him, "I'm going to arrest you if you don't move back!"

Meanwhile, I walked through the door. We've all seen photos of the stately door with the 10 on it. I walked through and I thought I would immediately be greeted. And I wasn't. There was no one there. Remember, this was early in 2001, before the World Trade Center attacks, and security hadn't yet gone on high alert. But still, I expected to be met by someone as I entered the British version of our White House.

The entrance itself was nothing elaborate, more like a vestibule. It was very proper and neat, but nothing ornate. As I stood there waiting to be greeted, I was listening for voices, but didn't hear any. I could've gone left, right or straight. I did a right turn. It became apparent in about six steps that I had gone in the wrong direction.

A rather large but proper gentleman stood in front of me and said in his polished British accent, "I think you need help in finding your way."

I said, "Yes sir, I do. I'm heading to a meeting with Prime Minister Blair."

He said, "Oh, you need to go the State Dining Room. Just head down that other hallway and up the stairs."

I did as he instructed, but again found myself all alone. There was no one else, no one to take me to the dining room. To this day, that amazes me!

When I found my way upstairs to the dining room, there were about 10 other journalists in the large room, which could have accommodated a much bigger crowd than what was assembled. There was other U.S. representation in the meeting. In those days, American news organizations had large bureaus with staff in many cities.

Prime Minister Blair was very candid in his conversation with us. He was warm, friendly with a great sense of humor. He was someone you could imagine having a beer with in a pub. It just felt like you had known him for a while, no barriers between him and you. While he was obviously very knowledgeable about the subject, he wasn't overbearing. He said, "We want your people to come back and see us. We need you to come here and visit us. We're such good friends and we want to maintain the friendship. Our countryside is open to you." I didn't get a one-on-one interview and I don't think anyone else did, either, but it felt like we had because there were

10 Downing Street was originally one of fifteen homes built on marshy ground by Sir George Downing, an Irish native who was raised in New England and was one of the first graduates of Harvard. Downing didn't do a particularly good job when it came to building on the soft soil. Winston Churchill wrote that Number 10 was "shaky and lightly built by the profiteering contractor whose name they bear." [1]

so few of us in the room.

Despite Blair's demeanor and appeal for calm, I was as concerned as ever that we in the United States needed to be prepared. Our livestock industry could have been ruined if Hoof-and-Mouth Disease had spread into this country. Our exports would have been immediately shut off, and we would have had to kill livestock by the tens, maybe hundreds, of thousands.

Before that trip, I had a couple of pork producers call me with concern about Hoof-and-Mouth Disease. They had been playing golf in Ireland, came back to the U.S. and headed straight to the hog farms where they lived. They and their bags had not been properly inspected by the USDA inspectors at O'Hare Field. The pork producers said they filled out the customs form which included the questions: *Have you been on a farm? Are you bringing agricultural products back?* These guys had very clearly checked "yes" on their forms, but there was no follow-up. They were taken aback that they were not detained and questioned about their whereabouts. To their credit, they had left their shoes behind. (I took their cue and also piled my shoes in a wastebasket before our flight home. I did not want to take any chance of bringing something back.)

During some of my farm reports on WGN Radio, I talked about

what happened to the pork producers, and that they felt the inspections were lacking at O'Hare. Well, WGN has a big audience and the word spread and wheels started turning. O'Hare brought in more inspectors and resources to adequately inspect at that time. I don't know if I had a role in that or not, but something had shifted after I talked about it.

Later, when Phil and I came back from London, our plane pulled up to the gate at O'Hare and before anyone deplaned, someone stepped on board. Shortly, there was an announcement: "Arriving passenger Armstrong, please report to the USDA inspectors immediately." Well, they had the dogs sniffing me and everything I owned. That was on a weekend and the man in charge was not there, but the word was out that they needed to make sure that "arriving passenger Armstrong" was thoroughly inspected. It was an hour and a half before I got out of O'Hare!

Whether that was done as a message for me or not, it was a reminder of how important it is to safeguard our food supply. The fear of Hoof-and-Mouth Disease remains realistic, and all kinds of scenarios and "what-ifs" have been developed. State agriculture departments all over the country have prepared for something of this nature. If we did have an outbreak of a contagious disease that could threaten our food supply, these

In the first five years after Tony Blair became prime minister in 1997, 37 computers, four mobile phones, two cameras, a mini-disc player, a video recorder, four printers, two projectors and a bicycle were stolen from 10 Downing Street. In 2015, it is one of the most heavily guarded buildings in Britain. No one can enter without passing through a scanner and a series of security gates manned by armed guards.[2]

various levels of federal, state and local governments have a plan. They know what to do.

The first responders would be veterinarians — the people who care for our cats and dogs as well hogs, cattle and other livestock. Federal veterinarians are stationed across the country and would be our first line of defense, swinging into action similar to the National Guard or our military. Inspection today is as important, if not more important, than ever, especially if someone wanted to intentionally do damage to our country. I feel that, as a nation, we are as prepared as we can be in safeguarding our food supply, but we've seen how easily things can sneak in. Invasive species from other countries can threaten our crops and our livestock; problems that other countries have and we don't want here.

All I have to do to realize that is look out my window here in North Carolina, where kudzu covers nearly everything along the road. Although kudzu was introduced with good intentions — in the 1930s and '40s, southern farmers were paid about eight dollars an hour to plant the vine as a way to prevent soil erosion — it's become an invasive species that threatens us in one way or another.

Bottom line: When you come back from another nation, don't lie about whether you've been on a farm, or that you're carrying a banana in your bag. For goodness sake, there's good reason they're asking you.

[1, 2] *The Daily Telegraph,* London, May 29, 2012

— *Chapter Thirty-One* —

Clayton Yeutter

nother of the interesting people I met in the Heartland is Clayton Yeutter, who grew up on a farm near Eustis, Nebraska. I've often told people that Clayton is the type of guy who can look you in the eye, tell you you're an idiot, and make you feel good about it. That honesty and candor are two of the reasons he has been such a great negotiator on behalf of the American farmer.

We met when he ran the Chicago Mercantile Exchange. He was the president of the Merc from from 1978-1985, before the merger with the Chicago Board of Trade.

Almost no one will remember him as being a Secretary of Agriculture. He was appointed by George H.W. Bush and served for the first two years of 41's first term before resigning to become chairman of the Republican National Committee. But where he really made a difference for the farmers and ranchers in the U.S. was as Special Trade Representative, and was referred to as Ambassador. He pried open the door into Japan for our meat. Although Japan had been a good grain customer of ours through the years, it had imported very little of our meat. With a population of 20 million and the space the size of California, it had limited space to produce and needed imports. But Japan aggressively protected its farmers from imports.

Then Clayton Yeutter started negotiating with them. He did so with such skill, not only there, but with the rest of the world, that today, exports account for about 25 percent of U.S. pork production, 17 percent of poultry production and 10 percent of beef production. Clayton's contributions as Ag Secretary may be overlooked, but he should never be forgotten for what he did as Special Trade Rep. I admire him a lot. He was honored in 2015 with an endowed chair at the University of Nebraska.

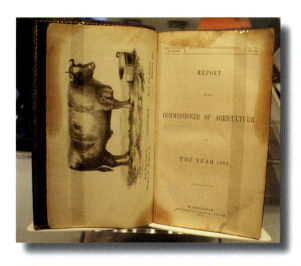

PRESIDENT LINCOLN CREATED the U.S. Department of Agriculture over 150 years ago. His personal copy of the first annual report from the Commissioner of Agriculture in 1862 is part of the exhibit at the Abraham Lincoln Presidential Library and Museum in Springfield, Illinois. Linda and I donated this book (with "Abraham Lincoln" embossed on the cover) to the museum in 2000.

The book had been given to us by a retired corporate executive who was a long-time, faithful WGN Radio listener. He called me out of the blue and asked me to come to his home. He said he and his wife were leaving the area and they wanted me to have this historic book, fearing that their heirs "would just put it in a garage sale someday."

The Lincoln Library and Museum is a GREAT place to visit. Go there if you haven't, and see more at www.presidentlincoln.org.

When President Reagan visited an Illinois farm, in October 1982, about a year-and-a-half after the attempt on his life, he rode in and drove the 8630 belonging to Chapin farmer John Werries. John still has the tractor and the memories of the day "The Gipper" went along for a ride. John told us that all of the panes of glass in the cab except the one on the far side, were removed for the television news coverage

John told me that just out of sight in this shot, right behind the cab, was Secret Service Agent Tim McCarthy. Not many months earlier, McCarthy (who went on to become the police chief in Orland Park, Illinois) had blocked the attempt on the President's life.

THIS RICHARD NIXON BUMPER STICKER had been hanging in the farm machine shed since the campaign of 1968, I do believe. Dad must have torn it, pulling it from the back bumper of the Pontiac Catalina. Or maybe it got torn five years later when Nixon resigned in the face of possible impeachment.

With the hindsight of 35+ years, and against the backdrop of the nation's current multiple crises, those times sure look tame, though we know they were not.

I HAVE ALWAYS THOUGHT a lot of President George H.W. since this day in Washington those many years ago. The subject was "agriculture." I was the moderator, and 21 television cameras were lined up on the riser in front of us.

I've never met anyone more genuine. When he spoke with you, he looked you in the eye ... and he also listened.

The "before" and "after" photos of Jim Armstrong's old Farmall 560. A proposed makeover on the bub in bibs was deemed a waste of time.

— Chapter Thirty-Two —

Old Iron

What is it about these old tractors? Why do grown men and women so thoroughly enjoy driving these old machines down the road in the company of other fans of vintage farm equipment? Why do we gather, by the tens of thousands, at events such as the Half Century of Progress Show in Rantoul, Illinois, and spend a few days telling stories and comparing with other tractor lovers? Or, how about the endless sharing of photos and stories on the Internet and social media?

Red tractors, green ones, along with orange, blue and gray ones, as well as some you just don't see that often. Old Iron, some people call them, but, oh, they look so much different now than when they were made. Many of them look better than when they rolled off the assembly line five, six, seven or more decades ago at places like Waterloo, Rock Island, Charles City, or Racine. With sandblasting, glossy paints and superbly remanufactured parts, these tractors truly are "standing tall."

Then there are some that just have not yet been to the paint shed. They are wearing their "work clothes," and there is good reason for that. Some of these old machines truly do get worked yet today. That they are still so functional is a testament to the magnificent engineering that went into these machines all those years ago.

And some might suggest that the finish on that steel is nothing to get too worked up about, anyway. While a mirror reflection in that paint sure is something to be proud of, we might hasten to point out that these tractors have, in their DNA, dirt, mud and manure, chaff and weed pollen, grease smudges, oil puddles and gasoline spills, and scratches and fading.

What hasn't faded is our emotional connection to those tractors. It may have been Dad's or Grandpa's. It may be just like the one Uncle Frank had. It could be the very same tractor we started farming with many, many tractors ago. Sitting on that seat brings back a flood of family memories. There are recollections of the changing seasons ... births and deaths ... first dates and graduations ... plowing, planting, cultivating, spraying, mowing and harvesting ... sweating and freezing ... and dreaming and scheming.

I won't be guilty, though, of romanticizing the memories. The days of working these tractors across those fields were very long. Work started early, and with many acres to cover before dusk, there was no time to waste. Hour after hour, we looked back to check the results of that moldboard plow as it turned over the soil behind the tractor. We watched the Kewanee disc slice through the plowed ground, preparing that seedbed for the planter. A few weeks later, we stared down at those rows of corn as they passed between the cultivator shovels, mile after mile and hour after hour. Oh, it was often drudgery, for sure, but from that tractor seat we had ample opportunity to contemplate the vast world beyond our farms, and what experiences may lie ahead of us. We had no clue of what was yet to come, with no idea that decades later, those long days and those experiences with these tractors would come to mean so much. And still, every time I settle onto the seats of those old, but shiny new, Farmstrong Farmalls, the memories come flooding back.

— Chapter Thirty-Three —

Heritage Tractor Adventure

*T*he idea for the Heritage Tractor Adventure came from Iowa, where the late Mark Pearson of WHO-AM Radio got the inspiration for the "WHO Radio Great Iowa Tractor Ride" from a bicycle ride. He and fellow broadcaster Gary Wergin figured that if people would ride their bicycles across the state in the annual RAGBRAI (Register's Annual Great Bicycle Ride Across Iowa), maybe farmers would congregate and drive their tractors long distances, too. They were right. It's been done for 19 years with 550 tractors participating in the most recent three-day event.

I wanted to see if we could do something similar in Illinois, even though we're in a more congested space. We worked with the Heritage Corridor Convention and Visitors Bureau, which is located in Will County and handles tourism for the communities and counties along the historic 97-mile Illinois and Michigan Canal National Heritage Corridor. The Heritage Corridor spans from the Chicago Portage Area just southwest of Chicago to LaSalle-Peru including the counties of Will, Grundy, LaSalle and Putnam.

We started it in 2000 and did it for about 13 years with a few hundred tractors joining in each year, and what I'm most proud of is that we did safely. Nobody got hurt. I wanted police agencies and municipalities all along the way to be involved, and they became valuable partners, providing escorts and other help. There was no passing and no clowning around. I emphasized that these tractors were unstable enough with just the three wheel touch points, and you had to be careful with them all the time.

I had a close call in downtown Joliet one year. We were sitting with our engines idling and my tractor was about six feet behind a squad car with the rest of the tractors behind me. I had my tractor in gear with my foot on the clutch. A woman holding a baby came up on the right side of the tractor. She looked up at me and said, "I'd like to take your picture holding little Leroy." Well, she heaved little Leroy up and over the right rear tire at me! As I grabbed the child, my left foot came off the clutch and the tractor lurched forward. I quickly depressed the clutch and the brake, but my tractor came within two feet of crashing into the back of the squad car. It was a great lesson for me about taking the tractor out of gear. Little Leroy will probably never know what a close call he had.

We met a lot of America's best people and it was great to showcase the ag heritage as we chugged through communities. One thing I really enjoyed, was doing loops past nursing homes along the way. The nursing homes had been alerted, and residents — most of them former farmers and farm wives — were gathered outside to look at the tractors. It was a special thing to see them reacting to the tractors going by, maybe bringing back some warm memories as well as a little bit of sunshine to their day!

We featured the annual rides on our TV show, and now there are tractor rides all over the U.S.

Everything looks better from the seat of a tractor. The Heritage Tractor Adventure was great annual fun. Tractors had to be at least 30 years old, have rubber tires, and be able to go at least six miles per hour. We are most grateful to Mary Beth DeGrush for the great effort she put into the ride each year. She made it happen!

The 11th annual Heritage Tractor Adventure in 2013 turned out to be the last one. It attracted tractors boys, and girls, of all ages and tractor colors and sizes. Great fun!

We had a lot of fun along the way, and also managed to do some good. I somehow wound up behind bars in Gardner, Illinois, with the longtime mayor, the late Tom Wise. The $640 in bail I needed to raise from friends for my freedom went to the Gardner area food pantries.

At the end of long, tiring days, we spent the nights in many great communities who welcomed us. At right, some of the 300 or so tractors that converged on Wenona, Illinois in 2013.

Chad Colby

The Farmstrong Farmall 560 and Darius Harms' Model 82 combine as pictured in Lee Klancher's beautiful photo book, Red Combines 1915-2015, *which is available at http://octanepress.com/book/red-combines-1915-2015*

This is an extremely rare 7288 owned by my friend Darius Harms. Only 19 of these were built by IH. This was photographed on Darius' family farm near Rantoul, Illinois, and appears in Lee Klancher's beautiful and award-winning book, Red Tractors 1958–2013. You can order it at http://octanepress.com/book/red-tractors.

HERE'S SOME "Old Iron" I wish I still had. Why in the world did we get rid of that '65 GTO?!

You can see a little dust across the back of the trunk lid. Think maybe the boy had a little gravel in his travel?

ONE OF agriculture's best friends in high places over the decades has been Mike Johanns. When this photo was taken, he was a U.S. Senator from Nebraska (2009-2015). He was Secretary of Agriculture (2005-07), and was governor of Nebraska (1999-2005).

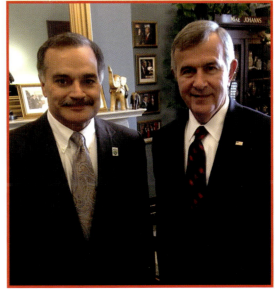

AS A YOUNG MAN, I would sit at high school basketball games with my mom and dad while my brother Steve was out on the floor. I had my transistor radio with me and would listen to Rich Lankford and his brother Ray Jay Lankford do the play-by-play. I was more interested in what those two guys were doing courtside than the actual game!

Rich, on the right, was 83 when this photo was taken and was in his 60th season of broadcasting Indiana high school basketball on WRAY at Princeton, Indiana. His broadcast partner is his son, Jeff.

The Lankford Family of fine broadcasters included Rich and Ray Jay's brother Stuart and his son Kent at WAKO, Lawrenceville, Illinois. And my friend Steve Lankford, Rich's nephew, is owner of WRAY-AM and WRAY-FM. Both stations are affiliates of *Max Armstrong's Midwest Digest*, I'm very proud to say.

Rich passed away in 2014 at the age of 85. I should have had Rich make a list of the kids whose games he covered before those guys went on to become star athletes. There were several, like a state trooper's kid from Heritage Hills High School named Jay Cutler.

This is the young man responsible for naming his <u>steers</u> "Orion" and "Max" in August of 2012 at the Mercer County, Ohio, fair. Seth Houts, 17, was a member of the Parkway FFA. He won Senior Showmanship and Senior Herdsman awards. Seth went on to study at The Ohio State University.

As for "Orion" and "Max," their futures weren't quite as bright.

Ralph W. Sanders

Chapter Thirty-Four

The "Super H"

*O*ne of my favorite shots of Jim Armstrong's 1953 Farmall "Super H" was taken early on a Sunday morning in June 1997. It was two summers after we hauled Dad's tractor north from Gibson County, Indiana, where it had been used on the farm for 42 years.

Iowa photographer Ralph Sanders met us at the taxi turnaround on Wacker Drive just east of the Hyatt Hotel in downtown Chicago. We spent about two hours photographing there and along Michigan Avenue.

In the background — as the camera looks to the northwest — left-to-right, is the old Sun-Times Building, the Wrigley Building, the Equitable Building (the last downtown location of IH headquarters) and Tribune Tower.

Perhaps you can imagine the thrill of driving this International Harvester tractor down Wacker Drive and across the Michigan Avenue Bridge next to where Cyrus McCormick had his reaper plant a century before.

Even on a quiet Sunday morning — quiet by Chicago standards — the "H" attracted plenty of attention, including, as you'll see on the next page, that of Chicago's finest.

"No, officer, you don't need to get your ticket book. I just took a wrong turn at Princeton. The next sign I saw said 'Dan Ryan Expressway.'" "Yes, sir, I know I can't stay parked in front of the Wrigley Building. Is there anything I can do to convince you to not write me a ticket? Of course, sir, come on up!"

FOR YOU IH-FARMALL FANS, the skyscraper seen here is the Equitable Building. Peeking into the frame from the left with a flag flying at its peak is the Tribune Building, where my office was for 23 years.

Cyrus McCormick built his reaper plant here in 1849. It was replaced in 1965 with this building, in which IH occupied 12 floors.

As late as 2005, Orion and I would see Brooks McCormick, the great-grandnephew of Cyrus, and the CEO of IH for many years, having lunch across Michigan Avenue at The 410 Club. We would speak with him briefly, but he liked his privacy. Brooks' father, Chauncey, was a cousin of Robert "Colonel" McCormick, the longtime publisher of the Trib.

As I recall, Brooks was still putting on a suit and tie and reporting to his office downtown up until about a week before he died in 2006. Often, as I came out of the Tribune Building to go to my pick-up truck in the NBC Tower lot, I would see his driver waiting on Lower Michigan. His Illinois license plate was 65. In later years, Brooks was a philanthropist and supporter of equine therapy.

The old McCormick horse farm near Warrenville was deeded over to the Forest Preserve District of DuPage County in 2000, with the stipulation that it would take possession of the farm upon McCormick's death, with the condition that several structures on the property, including McCormick's own mansion, be demolished.

Baling wire on the old "Super H" drawbar is apparently a necessity. Jim Armstrong always had it there, and I am not sure it would run without it!

During our last lap through the Rock Island Farmall Plant in the winter of 2005, photographer Mike Hood lit the scene superbly, and the red paint on the "Super M" really came to life.

The wrecking ball arrived not many weeks later. Several of the best minds in the I.H. Collectors Club Chapter 10 tried to in vain to save the mural. There was just no practical way to do it.

At left, top, is the M&W throttle on my "Super M."

The other photo shows the "M" of M&W, Elmo Meiners. He's the man with the hat and the smile, presenting a trophy to the winner of a tractor pull. It was at the one of the annual M&W Power Shows that were held between 1953 and 1973.

That young man wearing sunglasses and holding the microphone looks familiar.

Sherry Schaefer, Heritage Iron

THE M&W THROTTLE IS JUST one of many M&W after-market parts that made life easier for IH tractor owners. It started in 1946, when Elmo Meiners and Art Warsaw came up with a solution for the lack of speed ranges between the 4th and 5th gears on the Farmall. They worked in their spare time in the basement of Elmo's grain elevator in Anchor, Illinois, and developed a bolt-on, auxiliary transmission kit, the "nine-speed transmission," which became a big hit with "M" and "H" owners, and was the invention that got the M&W company going.

Before Elmo passed away at the age of 95 in 2009, he attended a Half Century of Progress Show in Rantoul, Illinois, and we paid tribute to him and his great work. On behalf of all IH owners whose days went better thanks to M&W parts, God bless you, Elmo Meiners.

Two or three times a year, towns, businesses or associations hire me to bring my tractor(s) to a community for a parade. I always enjoy it and am very appreciative of the kindness shown me. And I couldn't do this were it not for the superb care and transportation of my tractors by Ben Eipers.

My goal in the parade line-up is to be away from the fire truck sirens, not too close to the brass of the superb bands, in front of the horses, and a long distance from the politicians. I think you can understand the reasons for all four preferences.

In the photo above that I snapped from my tractor seat in Oregon, Illinois, I had to keep my distance from a fire-breathing Captain America. And I apologize for my response to the sweet, little, older lady who yelled from her wheelchair, "Hey, Max! Why don't you try that?"

ONE OF THE GREAT PERKS of my job is "having" to attend a number of county and state fairs, and one of the best is the Sandwich Fair west of Chicago. It's actually the DeKalb County Fair at Sandwich, Illinois, but is commonly called the Sandwich Fair. Since 1962, it has featured Fay's Pork Chop Bar-B-Que. The aroma from their grills pulls like a magnet.

Wilder Fay started the outdoor catering service as a way to add value to the hogs he and his family raised at Waterman, Illinois. Wilder and Martha had eight kids, most of whom work in the business today along with some spouses, kids and grandkids.

They are wonderful, friendly folks, including son-in-law Bob Dempsey, Marcia's husband, who took over as manager of Fay's upon the sudden death of Wilder in 1972.

Want More Max?

@maxarmstrong

Max Armstrong
@maxarmstrong FOLLOWS YOU

This Week In AgriBusiness & Max
Armstrong's Midwest Digest

📍 IL-IN-NC U.S.A.
🔗 farmprogressamerica.com
🕐 Joined February 2009

Tweet to   Message

Max Armstrong

Timeline   About

facebook.com/maxarmstrong

facebook.com/maxarmstrong.farmalltractors

broadcastermaxarmstrong

Watch Max and Orion on *This Week in AgriBusiness* on RFD-TV, many local stations, and on line at agbizweek.com. Hear Max on radio stations across the country with weekday broadcasts of his ag perspectives on *Farm Progress America*, his wit, wisdom and observations on *Max Armstrong's Midwest Digest*, and with Orion on *The Saturday Morning Show* on WGN Radio.

Download
Max's Tractor App

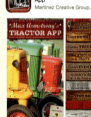

Max Armstrong's Tractor App
Martinez Creative Group, Inc.